A NATURALIST'S GUIDE TO THE

MAMMALS
OF
INDIA

Bangladesh, Bhutan, Nepal, Pakistan and Sri Lanka

Bikram Grewal and Rohit Chakravarty

JOHN BEAUFOY PUBLISHING

This edition first published in the United Kingdom in 2022 by John Beaufoy Publishing Ltd
11 Blenheim Court, 316 Woodstock Road, Oxford OX2 7NS, England
www.johnbeaufoy.com

10 9 8 7 6 5 4 3 2 1

Photo Credits
Front cover: *main* Tiger (Gopinath Kollur); *bottom left* Nilgai (Gopinath Kollur); *bottom centre* Golden Langur (Gopinath Kollur); *bottom right* Indian Fox (Clement M Francis).
Back cover: Black Panther and Leopard (Prakash Ramakrishnan). **Title page:** Malabar Giant Squirrel (Manoj Kejriwal).
Contents page: Hog Deer (Clement M Francis).
Main descriptions: photos are denoted by a page number followed by T (top), B (bottom), L (left), R (right), a (above), b (below).
Abhishek Jamalabad: 155B; **Aditya Singh:** 83B, 100T; **Amano Samarpan:** 52T, 53T; **Amit Sharma:** 62B, 64B, 80B; **Anand P:** 121T; **Anuroop Krishnan:** 113T; **Anwaruddin Choudhury:** 83T; **Aparajita Dutta:** 99T; **Ashok K Das:** 110B; **Ashok Kashyap:** 32T, 86B; **Ashwin H.P:** 21B, 23T, 28T, 39T, 56B, 87T, 88Br, 89Bl, 119T, 127B, 139B, 140T, 141T, 148T, 152B; **Ayan Banerjee:** 130T; **Bhavita Toliya:** 67; **Bikram Grewal:** 50T, 103B, 106T; **Biswapriya Rahut:** 104T, 104B; **Biswaroop Satpati:** 61Tr, 117B, 123T; **Clement M Francis:** 10, 22B, 23B, 30T, 33T, 39B, 41T, 45B, 79T, 106B; **Dhanu Paran:** 68B, 93B; **Dibyendu Ash:** 109T; **Dilan Mandanna:** 125T; **Divya Panicker:** 156B; **Dushyant Parasher:** 60, 61B, 71T; **Garima Bhatia:** 94Bb; **Giri Cavale:** 78T, 111B; **Gopinath Kollur:** 11T, 17, 30B, 36T, 44T, 47T, 54B, 55, 57B, 59, 65T, 73T, 73B, 74B, 85Bl, 91B, 96B, 97Tb, 99B, 114B; **HRS Urs:** 93T; **Kalyan Varma:** 89T; **Kamal Purohit:** 38T, 100B; **Karthikeyan Srinivasan:** 20T, 22T, 112B; **Khushboo Sharma:** 137T; **Manjula Mathur:** 48T, 49, 97Ta; **Manoj Kejriwal:** 28Bl, 68T; **N.A. Naseer:** 26T, 90T, 101B; **N.S. Dhingra:** 74T; **Nandini Velho:** 82T; **Narayan Sharma:** 25T; **Niazul H. Khan:** 98T; **Nikhil Devasar:** 34B, 54T, 76T; **Niranjan Sant:** 77B; **Nishma Dahal:** 108T, 108B; **Otto Pfister:** 80T, 81T, 81B; **Pallavi Ghaskadbi:** 154B; **Rajesh Puttaswamaiah:** 118T, 132T, 132B, 134B, 137T, 138B, 143T, 147T; **Rama Neelamegam:** 156T; **Ramki Sreenivasan:** 43, 51T, 107B; **Ranjan Kumar Das:** 112T; **Rohan Pandit:** 65Bl, 82B, 115T, 115B, 119B; **Rohit Chakravarty:** 29B, 34T, 69, 92T, 124T, 124B, 133T, 133B, 134T, 135T, 135B, 136T, 136B, 138T, 140Br, 141B, 142T, 142B, 143B, 144T, 144B, 145T, 145B, 147B, 148B, 149T, 149B, 150T, 150B, 151T, 151B, 152T, 153T, 153B; **Roon Bhuyan:** 21T, 24T, 36B, 102T, 120T, 118B, 120T, 120B; **Sachin Rai:** 105T, 113B; **Samyak Kaninde:** 63T, 64T, 64B, 94Ba; **Sarwandeep Singh:** 24B, 31T, 35, 42T, 78, 84B, 101Tr, 107T, 109B, 116T; Savio Fonseca: 117T, 129T; **Sayam U. Chowdhury:** 102B; **Soumyajit Nandy:** 105B; **Subrat Debata:** 137B; **Sujan Chatterjee:** 72T; **Sumathi Shekar:** 111T; **Sumit Sen:** 92B, 97B, 130B; **Taksh Sangwan:** 114T; **Tatiana Petrova:** 26B, 28Br, 29T, 54T, 62T, 84T, 103T, 125B, 131B; **Tharaka Kusuminda:** 129B, 139T; **Tripta Sood:** 20BL, 20Br, 27, 32B, 33B, 37B, 40B, 41B, 44B, 45T, 46B, 47B, 48B, 52B, 53B, 57T, 75T, 75B, 76B, 77T, 85Br, 86T, 91T, 94T, 95T, 95Ba, 95Bb, 96Bb, 101Tl, 106B, 126B; **Udayan Borthakur:** 25B, 121B, 122T, 122B, 154T; **Urmil Jhaveri:** 85T; Vaidehi Gunjal: 123B; **Vardhan Patankar:** 159B; **Vijay Kurhade:** 96T; **Vijay Sardesai:** 110T; **Vivek Sinha:** 38B, 40T, 42T, 46T, 56T, 58T.

The presentation of material in this publication and the geographical designations employed do not imply the expression of any opinion whatsoever on the part of the Publisher concerning the legal status of any country, territory or area, or concerning the delimitation of its frontiers or boundaries.

ISBN 978-1-913679-20-0

Edited by Krystyna Mayer
Designed by Gulmohur India

Printed and bound in Malaysia by Times Offset (M) Sdn. Bhd.

·Contents·

Introduction 4

Habitat 8

Classification 12

Glossary 18

Species Accounts 20

Checklist 160

Further Reading 173

Index 174

Introduction

The Indian region is incredibly rich in wildlife. This abundance is due to the variety of habitats and climate. Altitude ranges from sea level to the peaks of the Himalaya, the world's highest mountain range; rainfall from its lowest in the Rajasthan desert to the highest in the north-eastern town of Cherapunji in Meghalaya, one of the wettest places in the world. Unlike in more temperate zones, the climate of large areas of the Indian region encourages continuous plant growth, thereby providing an abundant source of nourishment.

The people of the region lived for thousands of years in close contact with its rich natural life. The earliest Hindu religious work, the *Rig Veda*, refers to several species of animal. Legends and myths grew around certain familiar species. Most Hindu gods and goddesses had mounts called *vahanas*.

The relationship between the deities and their mounts is depicted in both mythology and iconography. The mighty goddess Durga bestrode a Tiger when she slayed the demon buffalo. Ganesha, the elephant-headed god, has a mouse as his vehicle, while Yama, the god of death, rides a buffalo. The Moon god Chandra rides a Blackbuck. Almost every Hindu god is associated with a mammal, bird or even reptile.

The serious study of nature truly started with the arrival of the Great Mughals. They maintained royal menageries and hunted on a grand scale. But they were also meticulous in their observations of natural life.

The Emperor Babur, founder of the Mughal dynasty, was a man of many talents, and according to an observer had 'an undiminished interest in natural history, and a singular quickness of observation and accurate commemoration of statistical details'.

Goddess Durga with her tiger vahana

The elephant-headed god Ganesha on his mouse vahana

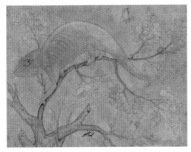

Chameleon painted in the court of Jahangir

Emperor Jahangir

His grandson Emperor Akbar, the mightiest of all the Mughal rulers, maintained more than 500 cheetahs, which he used to hunt game. He also had a huge retinue of court painters, who recorded the natural wonders that surrounded them.

His son Emperor Jahangir took his immense passion to an another level. He encouraged his court painters to develop nature studies and to present individual mammals, birds and

Cheetah hunt during reign of Emperor Akbar

plants to a level never attempted before. He showed great excitement when a new specimen was presented to him, and personally made copious notes and instructed his court painters to painstakingly record them.

The Mughal empire started to disintegrate at the beginning of the 18th century, and the spread of British power gave enormous scope to officers in the police, and civil, forest and armed services to observe the region's plentiful wildlife. As British power grew, there was an increasing number of such officers, whose jobs required much less crippling paperwork than those of their successors today. The result was the pioneering work of men such as Brian Houghton Hodgson, T. C. Jerdon and Edward Blyth.

The early study of mammals, like birds, was largely driven by hunters and restricted to trophy hunting. Richard Lydekker, employed by the Geological Survey in India, worked on vertebrate paleontology and wrote several books on Indian animals, including *The Wild Animals of India* and *The Game Animals of India*. Robert Armitage Sterndale was another early pioneer. Originally an indigo planter in Bihar, he later went on to become the Governor of St Helena. In 1884 he published the path-breaking *Natural History of the Mammals of India and Ceylon*.

Robert Charles Wroughton, an officer of the Indian Forest Service, conducted the first collaborative mammal survey from 1911 until 1923. Assisted chiefly by C. A. Crump, J. M. D. Mackenzie, Capt. Philip Grosse, Sir Ernest Hotson, R. Shunkara Narayan, S. H.

Samuel Tickell

Richard Lydekker

Brian Hodgson

T. C. Jerdon

Robert Sterndale

A. O. Hume

Edward Blyth

S. H. Prater

The Tiger *captured the imagination of the British*

Prater and Charles McCann, this survey collected more than 60,000 specimens, especially small mammals, and the information was published in 47 papers. Several new specimens were also described.

Brian Houghton Hodgson, a civil servant, served as the British Resident in Kathmandu for several years. During his stay in Nepal he was never allowed to travel beyond the valley, so he employed several people to bring him specimens. He discovered the Tibetan Antelope, besides 39 other species of animal.

Edward Blyth was an English zoologist who spent most of his working life as a curator of the Asiatic Society of Bengal, in Calcutta. His onerous tasks prevented him from doing much field work, and he depended on specimens sent to him by other amateur collectors like Col Samuel Tickell, A. O. Hume and Robert Swinhoe. He was a friend of Charles Darwin and it is thought that he may have influenced his thinking.

The British Government in India undertook the massive exercise of publishing a major series of books under the general umbrella of *The Fauna of British India*. The two mammal volumes were published in 1886 and 1891, under the editorship of W. T. Blanford. The second revised editions, under the editorship of R. I. Pocock, were issued in 1939 and 1941. To this day, they remain a repository of detailed knowledge.

Stanley Henry Prater was an English naturalist who served as the curator of both the Bombay Natural History Society and the Prince of Wales Museum in Bombay. In 1948 he published *The Book of Indian Animals*, which remained the most popular book on Indian mammals for years, until many good works published by Indian naturalists appeared.

HABITAT

Oriental and Subregions

The Indian subregion is divided into seven distinct areas, where different types of animal are found. Portrayals of the significance of trees to these areas are consistent historically, across time and faiths, as is evident in various Mughal tombs and mosques, where the botanical patterns appear repeatedly.

The northernmost area of this region is the Himalaya, which forms an arc some 2,500km long and 150–400km broad across the top of the subcontinent. The Himalayan mountains form roughly three parts: the foothills of the Shivaliks to the south, the Himachal, or lower mountains, and the Himadri, or high mountains, to the north.

The Ladakh plateau, with an average elevation of 5,000m, occupies a large portion of the Indian state of Jammu and Kashmir. Several Himalayan species occur in this area, like the Bharal, Markhor and Ibex, and their shadowy predator – the almost mystical Snow Leopard. The Kiang, or Tibetan Wild Ass, is a much-sought resident of the Ladakh plateau.

The hill states of Himachal Pradesh and Uttarakhand lie to the west of Nepal, and fall almost entirely within the central Himalaya. Further east the increasing rainfall gives the Eastern Himalaya of Bhutan and Sikkim a very different range of species from those in the west. A prime example is the Golden Langur in Assam and Phayre's Leaf Monkey of Tripura.

The north-west covers the bulk of Pakistan, the flat plains of the Indian Punjab and

Himalayan mountains

Desert of Rajasthan

the semi-arid and arid plains of Rajasthan in the west. The Punjab (now divided between India and Pakistan) is watered by the five rivers after which it takes its name, and due to efficient farming on the fertile soil it produces an immense surplus of wheat and rice. Further west, wherever irrigation has been possible the desert has bloomed. Mountains of red chillies, for example, can been seen drying next to the fields around Jodhpur, while there are verdant paddy fields in areas irrigated by the great River Indus in Pakistan's Sindh province. Areas without irrigation have to rely on the perennially deficient rainfall, but local grasses have adapted to this, and after a monsoon shower even the desert sprouts rich pasture. Much of the area is, in fact, thorn scrub rather than true desert. Among the numerous mammals found here are the Blackbuck, Nilgai and Chinkara antelopes.

The increasingly rare Desert Cat, Striped Hyena and both the foxes – Indian and Bengal – are found here, as is a variety of jirds and gerbils.

The shifting sands of the desert join ultimately with the Rann of Kutch, a large salt waste that runs into the sea and is bordered to the south-east by the Aravallis, India's most ancient mountains. This is home to the Indian Wild Ass, now under severe stress due to the threat to its habitat. Another beleaguered species – the Indian Wolf – is hardly ever seen nowadays.

North India comprises the Gangetic plain, enriched by thousands of years of alluvial deposits brought by the River Ganga and its tributaries from the Himalaya. The Gangetic plain is densely populated and highly fertile. This region extends up to an altitude of 1,000m in the north, so that it includes the low foothills of the Himalaya, and the terai of India and Nepal, once a marshy area covered with dense forest. Much of the terai area has been cleared for farming, but some of the forests that still exist reveal the fantastic variety of animal life they must once have supported. The once-dense forests of Sal held large numbers of the mighty Indian Elephant. Unfortunately, the massive deforestation caused by human greed has lead to terrible human-animal conflicts in the region.

Peninsular India, bordered on the north-west by the Aravallis, the north by the Vindhya mountains, the west by the Arabian Sea and the east by the Bay of Bengal, makes up the largest physiographic division of India. The central plateau of this area, which is also known as the Deccan, rises to more than 1,000m in the south, but hardly exceeds 500m in the north. The peninsula has some wonderful landscapes, hills and huge boulders littering the countryside, and large areas of forest. Great rivers like the Narmada rise in the heart of the peninsula and flow into the sea. The steep escarpments of the Western Ghats, the mountains of which stand between the plateau and low-lying coastal strip, catch the full force of the monsoon.

The south-west region lies within the peninsula, with the highest of the hills here being the Nilgiris or Blue Mountains, much of whose characteristic downland and shola forest is now under eucalyptus, tea and other plantation crops. Tea is also the main crop of the Annamalai or Elephant Mountains of Kerala, while cardamom and other spices are grown lower down. The Nilgiri Tahr, Malabar Giant Squirrel and Nilgiri Langur are found here. Recent sightings of the Nilgiri Marten have brought some much-needed cheer to animals lovers of the area.

Forests of peninsular India

Forests of Central India

The north-east and Bangladesh region consists of the delta of the Ganga and Brahmaputra, with its tidal estuaries, sandbanks, mudflats, mangrove swamps and islands. Further upstream are lands drained, and occasionally flooded, by these great rivers and their tributaries. The north-east region also extends northwards to include all the forest regions of the states of Arunachal Pradesh, Mizoram, Meghalaya and Nagaland, as well as the Kingdom of Bhutan. As you progress eastwards, the birdlife has increasingly strong affinities with the Indo-Chinese subregion.

The Indian Rhinoceros is now restricted to this part of the world. The Hispid Hare and Pygmy Hog were thought to have become extinct, but fortuitously both were 'rediscovered' in the forest of the Manas National Park, adjoining Bhutan.

Sri Lanka, too, is a remarkable area for animal life. Although far from large, the country has a wide range of climates and habitats that support some amazing animals. The Togue Macaque is only found on this island. Sri Lanka can be divided into three zones: the dry plains of the north, the mountainous central region and the humid wet zone around the capital Colombo.

Wetlands of the north-east

CLASSIFICATION

ORDER PRIMATES

Primates are represented in South Asia by three families, seven genera and 30 species.

Lorisidae This family of small primates, commonly known as lorises, consists of three species, two of which are endemic to South Asia. The Slender Loris *Loris lydekkerianus* is found in Sri Lanka and South India. The Slow Loris *Nycticebus bengalensis* occurs in the north-eastern part of India. Both species are often hunted for food, and frequently kept as house pets.

Cercopithecidae The 27 species of Old World monkey in South Asia represent almost a quarter of the monkey species found in the world. Nine of them are macaques, and while the langurs have now been split into 18 different species, many are endemic to South Asia. The Toque Macaque *Macaca sinica* is restricted to Sri Lanka, while three species are confined to India. The Arunachal Macaque *M. munzala* was described as a new taxon as recently as 1997. As it name suggests, it is found only in the forests of Arunachal Pradesh and is considered highly endangered.

The langurs, as a family, are venerated in many parts of India, as they are considered as the chief of the Hindu god Lord Rama's army. They are therefore seldom killed or troubled. Originally treated as one species, with several subspecies, they have now been split into many distinct species, though some scientists are still debating these splits.

Hylobatidae Originally, the Hoolock Gibbon, India's only ape, was considered as one species, but it was recently split into the Eastern and Western Hoolock Gibbons. Mostly arboreal, they are hunted extensively for food by the various tribes of north-eastern India.

ORDER PROBOSCIDEA

Elephantidae The Asian Elephant *Elephas maximus indicus*, the largest land mammal in the region, has a wide distribution, especially in the northern, eastern and southern parts of the subcontinent. Its shrinking habitat, due to the demands of a burgeoning population, has resulted in almost daily conflict with humans. Elephants are often killed for their ivory.

ORDER PERISSODACTYLA

Six species of odd-toed ungulate occur in South Asia.

Rhinocerotidae The only representative found in the region is the Indian Rhinoceros

Rhinoceros unicornis. Historical records show that it was once present from Pakistan to Assam. However, it is now confined to small patches of grassland in east and north-eastern India. A few individuals have been introduced in the Dudwa National Park in the terai region of Uttar Pradesh. Hunted for its horns, which are in great demand as an aphrodisiac, the species is now critically endangered. Javan and Sumatran Rhinoceroses are no longer found in the region.

Equidae Only two species of equine occur in South Asia. The Asiatic Wild Ass *Equus hemionus* is confined to Gujarat, and the Tibetan Wild Ass *Equus kiang* is only found in the high plateau of Ladakh.

ORDER ARTIODACTYLA

There are 59 species of even-toed ungulate in South Asia.

Suidae Two species of pig occur in South Asia. The Wild Pig *Sus scrofa* is treated as vermin in some Indian states and is often killed by farmers. The Pygmy Hog *Porcula salvania* was thought to be extinct, but was rediscovered in Assam. A breeding programme has been established and it is being reintroduced in various forests of Assam.

Tragulidae The Chevrotain or Mouse-deer *Moschiola meminna* has recently been split into three species.

Moschidae 4 species of musk deer are found in different parts of the Himalaya. The males have canines instead of antlers and are hunted for their musk glands.

Cervidae 37 species of deer are found in the region. The Spotted Deer *Axis axis*, is the most common. The Swamp Deer *Rucervus duvaucelii* is found in India and Nepal. The newly found Leaf Deer *Muntiacus putaoensis* has now been seen in extreme North-east India.

Bovidae 37 species (including several recent splits) of bovine occur in the region. The Nilgiri Tahr *Nilgiritragus hylocrius* is confined to the Western Ghats in India.

ORDER CETACEA

Thirty-two cetacean species are said to occur in the waters in and around South Asia.

Balenopteridae Five species of rorqual are known to occur in and around the oceans of South Asia.

Delphinidae 17 species of marine dolphin are found in the region's oceans.

Phocoenidae The Indo-Pacific Finless Porpoise *Neophocaena phocaenoides* is the only representative of this family in the region's waters.

Physeteridae All three species of the world's sperm whales occur in South Asia's oceans.

Platanistidae Both species of the world's river dolphins are found in South Asia, and both are endemic to the region. While the Gangetic Dolphin *Platanista gangetica* is found in the rivers of India, Nepal, Bhutan and Bangladesh, the Indus River Dolphin *P. minor* is thought to be confined to Pakistan. However, there has been a purported sighting from Harike in Western Punjab.

Ziphidae Four species of beaked whale are found in South Asia's waters.

Kogiidae Both extant species of this family, Pygmy Sperm Whale *Kogia breviceps* and Dwarf Sperm Whale *K. sima* are found, albeit rarely, in South Asia.

ORDER CARNIVORA

Felidae South Asia has 18 species of cat, of which the diminutive Rusty Spotted Cat *Prionailurus rubiginosus* is endemic to the region. The Tiger *Panthera tigris*, which has a large, though thin distribution all over the region, barring Gujarat and Kashmir, has now been reduced to about 1,800 individuals from about 30,000 a century ago. The Asiatic Lion *Panthera leo persica* is now confined to a small region in Gujarat. The current population of about 350 is under constant threat, and all attempts to translocate it to other suitable habitats have proved fruitless. The Asiatic Cheetah *Acinonyx jubatus* is now extinct, the last three specimens having being shot in 1948. Plans for reintroducing it have been officially dropped.

Prionodontidae Only one species of this civet-like family is found in South Asia – the Spotted Linsang *Prionodon pardicolor*.

Viverridae 13 species of palm civet and civet cat occur in South Asia. Three are endemic to the region, of which two, the Malabar Civet *Viverra civettina* and Brown Palm Civet *Paradoxurus jerdoni*, are confined to the Western Ghats in South India, and the Golden Palm Civet *Paradoxurus zeylonensis* to Sri Lanka.

Herpestidae Seven of the world's 34 mongooses occur in South Asia. One species, the Marsh Mongoose *Herpestes palustris*, is endemic to India.

Hyaenidae Only the Striped Hyaena *Hyaena hyaena* is found in South Asia. It is rarely seen today.

Canidae Nine species of the dog family are found in the Indian subcontinent. The only species of canid that is endemic to the region is the Bengal Fox *Vulpes bengalensis*.

Ursidae Four species of the world's bears occur in South Asia. The widespread Sloth Bear *Melursus ursinus* is endemic to South Asia and is common in India, Sri Lanka and Nepal. The Red Panda *Ailurus fulgens* is also currently classified under this family.

Mustelidae 17 species of weasel, badger, otter and marten are found in the region. The rarely seen Nilgiri Marten *Martes gwatkinsii* is endemic to the region.

ORDER LAGOMORPHA

Ochotonidae Ten of the world's 30 mouse-hare and pika species are found South Asia.

Leporidae Of the world's 61 leporid (hares and rabbits) species, six are seen in South Asia. The Hispid Hare *Caprolagus hispidus*, sometimes referred to as the Assam Rabbit, was thought to be extinct until it was rediscovered in the Manas National Park in Assam.

ORDER PHOLIDOTA

Manidae Three species of pangolin are present in the region. Pangolins are hunted mercilessly in the region for their scales and other parts, which are in great demand for traditional medicine in Southeast Asia.

ORDER SCANDENTIA

Tupaiidae Three species of tree shrew are found in South Asia. Rarely studied, they and their distribution still remain unclear.

ORDER EULIPOTYPHLA

Erinaceidae Three species of hedgehog are endemic to the region. The Madras Hedgehog *Paraechinus nudiventris* is found only in South India, while the Collared Hedgehog *Hemiechinus collaris* and Indian Hedgehog *Paraechinus micropus* occur in India and Pakistan.

Soricidae 43 shrew species occur in South Asia, of which 18 are endemic to the region. In Sri Lanka the Singaraja Shrew *Crocidura hikmiya* was recently described.

Talpidae Four species of mole are found in the region.

ORDER RODENTIA

Sciuridae Thirty-nine species of squirrel are found in South Asia, of which 11 are endemic to the region.

Gliridae Only two species of the world's 28 dormouse species are found in South Asia.

Dipodidae Six of the 51 species of jerboas in the world occur in South Asia. One species, the Baluchistan Pygmy Jerboa *Salpingotulus michaeli*, is found in Pakistan and Afghanistan.

Platacanthomyidae The Malabar Spiny Dormouse *Platacanthomys lasiurus* is endemic to the Western Ghats in South India.

Sminithidae The Kashmir Birch Mouse *Sicista concolor* is the sole representative of this family in the region.

Spalacidae Four species of bamboo rat are seen in South Asia.

Calomyscidae Three out of eight species of mouse-like hamsters are seen in South Asia. The Hotson's Brush-tailed Mouse (*Calomyscus hotsoni*) is confined to Pakistan.

Cricetidae Of the 21 species of vole and hamster found in the region, the White-tailed Mountain Vole *Alticola albicaudus* and Central Kashmir Vole *A. montosa* are endemic to India, while the Royle Mountain Vole *A. roylei*, True's Vole *Hyperacrius fertilis* and Murree Vole *H. wynnei* are also found in Pakistan.

Muridae This large family of rodents is represented in the region by 84 species in 29 genera, including 29 that are endemic to the region, of which 13 are found in India. Sri Lanka has five while the Himalayan Field Mouse *Apodemus gurkha* is found in Nepal.

Hystricidae Three porcupines occur in the region. All are hunted for food, and their quills are illegally sold as tourist souvenirs.

ORDER CHIROPTERA

Bats comprise the second largest order of mammals, with 1,116 species in 202 genera all over the world, of which 152 species, in 50 genera, are found in South Asia.

Pteropodidae 16 species of fruit bat are found in the region. Salim Ali's Fruit Bat *Latidens salimalii* and the Nicobar Flying Fox *Pteropus faunulus* are endemic to India.

Rhinolophidae 25 species of horseshoe bat occur in South Asia. The Andaman Horseshoe

Bat *Rhinolophus cognatus* and Mitred Horseshoe Bat *R.mitratus* are endemic to India.

Hipposideridae 18 species of leaf-nosed bat are seen in the region. Four are endemic to the region, of which Khajuria's Leaf-nosed Bat *Hipposideros durgadasi* and Kolar Leaf-nosed Bat *H. hypophyllus* are seen only in India.

Mgadermatidae The false vampire bats are represented by the Greater False Vampire Bat *Megaderma lyra* and Lesser False Vampire Bat M. *spasma*.

Rhinopomatidae Three species of mouse-tailed bat occur in South Asia.

Emballonuridae The bats in this family are called 'sheath-tailed' or 'tomb bats'. Seven species are found in the region.

Molossidae Four species occur in the region. Wroughton's Free-tailed Bat *Otomops wroughtoni* was thought to be endemic to the Western Ghats, but has now also been reported from some caves in the north-eastern state of Meghalaya.

Vespertilionidae 75 species of evening bat occur in South Asia, of which 10 are endemic to the region. The Sombre Bat *Eptesicus tatei* and Peter's Tube-nosed Bat *Murina grisea* are found only in India. Escorba's Mouse-eared Bat *Myotis csorbai* is found only in Nepal.

Miniopteridae Three species occur in the region. The Eastern Bent-winged Bat *Miniopterus fuliginosus* (earlier considered synonymous with the widespread Schreiber's Bent-winged Bat) is the most common and widespread species in South Asia.

ORDER SIRENIA

Dugongidae The Indian subcontinent has only the Dugong *Dugong dugon*. The recent decision of the Indian government to deepen the Gulf of Mannar, which separates India and Sri Lanka, is causing great concern. Should this take place, the future of this rare mammal looks bleak.

Elephants of the terai

Glossary

aberrant Abnormal or unusual.
accidental Vagrant.
adult Mature; capable of breeding.
anterior Area on front part of body.
antlers Growth on top of head of a deer that grows annually.
aquatic Living in water.
arboreal Living in trees.
biodiversity Having many different life forms within an area.
blow Visible cloud of warm, moist air expelled by a whale while surfacing.
blubber Layer of fatty tissue below the skin.
burrow Hole in the ground dug by animals like hares.
canine Taxonomic group that includes dogs, wolves and foxes.
carnivore Meat-eating animal.
cetacean Of the mammalian order Cetacea, including whales, dolphins and porpoises.
crepuscular Active at dawn and dusk.
diagnostic Sufficient to identify a species.
diurnal Active during daytime.
duars Forested areas south of eastern Himalaya.
echolocation Navigation by sound.
endangered Facing high risk of extinction.
endemic Indigenous and confined to a place.
extinct No longer in existence.
family Specific group of genera.
feral Escaped and living in the wild.
foraging Searching for food.
form Subspecies.
genus Group of related species (plural genera).
Ghats Hills parallel to east and west coasts of India.
herbivore Animal that feeds on plants.
horns Growth on head of an antelope that never sheds.
insectivore Animal that feeds on insects.
juvenile Immature or pre-adult.
mangrove Coastal salt-resistent trees or bushes.
marine Able to live in salt water.
migration Seasonal movement between distant places.
monotypic Of a single form with no subspecies.
montane Pertaining to mountains.
nocturnal Active at night.
nomadic Species without specific territory.
nominate First subspecies to be formally named.
Old World From Asia and Africa.
order Group of related families.

organism Anything that is alive.
palearctic Large ecozone consisting of Europe and Asia above the Himalaya.
plankton Tiny organisms that float or swim weakly in the ocean.
poaching Illegal hunting.
predator Animal that feeds on other animals.
primates Members of a highly developed order of animals.
race Subspecies.
range Geographical area inhabited by a species.
resident Non-migratory and breeding in the same place.
rorqual Baleen Whales.
Sal Dominant tree of North Indian forests.
savanna Open flat land with grasses or small shrubs.
sholas Small forests in south-west Indian valleys.
species Group of animals reproductively isolated from other such groups.
submontane Hills below the highest mountains.
subspecies Distinct form that does not have specific status.
tank Water reservoir.
taxonomy Science of classifying organisms.
Teak Dominant tree of South Indian forests.
terai Alluvial stretch of land south of the Himalaya.
ungulate An animal that has hooves.
vagrant Accidental, irregular.
zoology The study of animals.

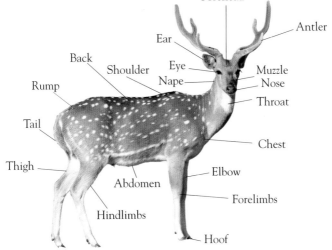

Western Hoolock Gibbon ■ *Hoolock hoolock*
F Hylobatidae **O** Primata

DESCRIPTION Body colour entirely black, with continuous white brow with curved ends. Distinctive dark brown genital tuft. Females lighter brown than males, with whitish ring around eyes. Males weigh about 6–8kg, females around 5–7kg. **DIET** Fruits, figs, leaves, insects and spiders. **DISTRIBUTION** Assam, Garo Hills in Meghalaya, Brahmaputra, Dibang and Lohit River Valley. **HABITAT** Lowland and hill forests. Tropical and subtropical wet evergreen forests.

Eastern Hoolock Gibbon ■ *Hoolock leuconedys*
F Hylobatidae **O** Primata

DESCRIPTION Male body colour entirely black, with separated white brow with curved ends. Distinctive white/rust-coloured genital tuft. Females lighter brown than males, with whitish ring around eyes. Hands lighter than body colour. Males weigh about 6–8kg, females around 5–7kg. **DIET** Fruits, figs, leaves, insects and spiders. **DISTRIBUTION** Lohit district of Arunachal Pradesh. **HABITAT** Lowland and hill forests. Tropical and subtropical wet evergreen forests.

Bengal Slow Loris
▪ *Nycticebus bengalensis*
F Lorisidae **O** Primata

DESCRIPTION Body weight about 2kg, and length around 25–35cm. Large eyes with brown circles around them. Stout body with silvery coat; head and shoulders greyish-cream. Distinctive brown stripe runs along back to crown. **DIET** Omnivorous, feeding on fruits and animal matter like birds' eggs, small birds, larvae of insects and similar. **DISTRIBUTION** Eastern India in Assam and parts of north-east. **HABITAT** Tropical and subtropical evergreen and semi-evergreen rainforests.

Slender Loris
▪ *Loris lydekkerianus*
F Lorisidae **O** Primata

DESCRIPTION Body weight variable, at 85–350g, and body length about 18–26cm. Second digits on both hands and feet are reduced. Body colour varies from dark grey to brown, with silverfish hairs in between. Underside paler than upperside. Dark spinal stripe less prominent than in previous species. Eyes encircled with black and white muzzle. **DIET** Omnivorous, feeding on berries and insects, lizards, small birds and frogs. **DISTRIBUTION** South India. **HABITAT** Dry deciduous and moist deciduous forests.

Rhesus Macaque ■ *Macaca mulatta*
F Cercopithecidae **O** Primata

DESCRIPTION Common macaque of northern India. Squat and sturdy with sandy-brown or tawny-red coat. Face essentially naked and pink. Differentiated from other macaques

by orange-red fur on loins and rump. Short tail, about 30cm in length. High variation in size and amount of fur, the largest animals with thickest hair generally being found in North India and Himalaya. Well habituated to human settlements, where it lives alongside humans, stealing food from houses and raiding crops and fields. Vocal, with a squeaky, raucous alarm call. **DIET** Omnivorous, eating insects and fruits in the wild, but thrives on all sorts of fruits, vegetables and human-mediated foodstuffs in habitations. **DISTRIBUTION** Himalaya, North India, North-east India and up to northern Andhra Pradesh in the peninsula. Range demarcated by Tapi River in west and Godavari in east. **HABITAT** Human habitation, agricultural fields, and tropical and temperate open woodland; rarely, if ever, in dense forest.

Bonnet Macaque ■ *Macaca radiata*
F Cercopithecidae **O** Primata

DESCRIPTION The common macaque in southern India. Similar in general body structure to Rhesus Macaque (see opposite), but differs primarily in two key features. Head has a distinctive parting of hair shown by no other macaque in the region. Also has long tail unlike short tail of Rhesus. Face naked and pink. Similar in behaviour to Rhesus. Vocal, with a squeaky, raucous alarm call. **DIET** Similar to that of Rhesus. **DISTRIBUTION** South India south of Tapi and Godavari Rivers. Ranges overlap with those of Rhesus around Mumbai and also in Godavari Basin. **HABITAT** Similar to that of Rhesus.

Assamese Macaque ■ *Macaca assamensis*
F Cercopithecidae **O** Primata

DESCRIPTION Similar in build and appearance to Rhesus Macaque (see opposite), with the same heavy and thickset build, but Assamese Macaque can be told apart by absence of orange-red patch on loins and rump. In some cases, even the habitat is indicative. Coat generally darker than that of Rhesus, with greyer rather than redder tones. Also often more woolly than Rhesus and occasionally bearded. In most localities, not as habituated to humans as Rhesus, keeping to forests in close-knit troops. **DIET** Primarily insects and fruits, but takes human-mediated food where habituated. **DISTRIBUTION** North Bengal, Sikkim and North-east India. Ranges overlap with those of Rhesus throughout its range. **HABITAT** Low-lying forests and montane broadleaved forests, but also edges of forests and villages; generally not accustomed to human habitation.

Arunachal Macaque ▪ *Macaca munzala*
F Cercopithecidae O Primata

DESCRIPTION Species that was recently described as separate from Assamese Macaque (see p. 23). Very similar to Assamese, but known to be slightly darker with a dark brown and unbearded face. Currently not reliably separated from Assamese by morphology. A little-known species. **DIET** Probably similar to that of Assamese. **DISTRIBUTION** Currently range restricted, with only a small known population near Tawang and West Kameng districts of Arunachal Pradesh. More studies are required to ascertain its true geographical range. **HABITAT** Subtropical and temperate montane forest edges near human settlements.

Long-tailed Macaque ▪ *Macaca fascicularis*
F Cercopithecidae O Primata

DESCRIPTION Medium-sized macaque with olive-brown coat and long tail. Olive-brown of coat often bordered with white that extends to face and delineates it. Often a short

crest on middle of head. Furry face and slight mane. Although arboreal, it procures most of its food on the ground in tidal mudflats. Lives in small troops of up to ten individuals. Lives close to human habitation and often picks food from houses. **DIET** Omnivorous, feeding on crabs, molluscs, insects and fruits. **DISTRIBUTION** In India, found only in Katchal, Camorta, Little Nicobar and Great Nicobar Islands. Common in South-east Asia, which forms the main range of its distribution. **HABITAT** Edges of tropical evergreen forests, mangroves, tidal mudflats, and coconut and *Areca* nut plantations.

Northern Pig-tailed Macaque ▪ *Macaca leonina*

F Cercopithecidae **O** Primata

DESCRIPTION Large, forest-dwelling primate, easily separated from other macaques in its range by squarish head and distinctive cap of short black hair. Tail moderate in size and thin, very similar to that of a pig (hence the name). While walking the tail is held erect. Fur usually light brown and small mane around face. **DIET** Insects and fruits, as well as human-mediated food in places where habituated. **DISTRIBUTION** Vulnerable species. North-east India south of Brahmaputra. **HABITAT** Moist lowland and montane broadleaved forests and forest edges; around villages where habituated.

Stump-tailed Macaque

▪ *Macaca arctoides*
F Cercopithecidae **O** Primata

DESCRIPTION Large and stocky monkey with distinctly bright and bald red face. Coat thick and furry, and bearded face. Tail, as the name suggests, reduced to stump and often invisible. Infants bright golden at first, gradually changing to white and finally brown, which is the adult pelage. **DIET** Insects and fruits. **DISTRIBUTION** Vulnerable species; shy and commonly seen only in certain pockets. North-east India south of Brahmaputra. **HABITAT** Moist lowland and montane broadleaved forests and forest edges; restricted to forests.

Lion-tailed Macaque ■ *Macaca silenus*
F Cercopithecidae **O** Primata

DESCRIPTION Medium-sized macaque, rather stocky in build with distinctive appearance. Immediately identifiable by dark, shaggy, jet-black coat and ashy-grey to white mane. Tail moderate in size and slightly bulbous at tip, like that of a lion. Chiefly arboreal, seldom descending to the ground to find food or to cross canopy breaks. Calls like a pigeon. **DIET** Fruits and insects, but there are also reports of it picking flying squirrels from their roosts in tree hollows. Partial to fruits of *Mesua ferrea*. **DISTRIBUTION** An endangered primate. Scattered and isolated locations in Western Ghats of Kerala, Tamil Nadu and southern Karnataka. Strongholds include Annamalai Wildlife Sanctuary and Valparai plateau, Silent Valley National Park and Someshwara Wildlife Sanctuary. **HABITAT** Dense and moist evergreen forests with continuous canopy cover.

Nilgiri Langur ■ *Semnopithecus johnii*
F Cercopithecidae **O** Primata

DESCRIPTION Large black langur with greyish or yellowish-white mane. Fur shaggy and black. Rump and base of tail often grizzled. Females are identified by white patch on inner sides of thighs. Young reddish for the first two months of their lives, then turn darker. Strictly arboreal and lives high up in trees. Vocal, and its booming *hoo* can be heard at dawn and dusk. **DIET** Herbivorous, eating leaves and fruits of various forest plants; has even adapted to feed on flowers and buds of wattles that have invaded its forests. **DISTRIBUTION** Scattered locations in Western Ghats south of Kodagu (Coorg). **HABITAT** Dense and moist evergreen forests with good canopy cover.

Northern Plains Langur ■ *Semnopithecus entellus*
F Cercopithecidae **O** Primata

DESCRIPTION Until recently this familiar langur was described as a single species, the Common Langur (*S. entellus*) composed of several subspecies. Recent studies have elevated most of the subspecies to species level. Subspecies differ mainly in amount of fur on body and carriage of tail. Northern Plains Langur is the most widespread of the langurs in the region. It is of medium size with a greyish-white coat and black face. Long, whip-like tail forwards-looped while walking. Mostly seen in troops of females or bachelor males, or harem of males with multiple females. Well adapted to human habitation. **DIET** Leaves, fruits, flowers and buds, but also feeds on human-mediated food. **DISTRIBUTION** North of Deccan Plateau, range demarcated by Ganges, Narmada and Godavari. Found from Gujarat in west to West Bengal in east, and Haryana to Telangana in the north–south axis. **HABITAT** Dry and open forests, and also near human habitation.

South-western Langur ■ *Semnopithecus hypoleucos*
F Cercopithecidae O Primata

DESCRIPTION Replaces Northern Plains Langur (see p. 27) south of Narmada. It is told apart from Northern Plains by rather dark coat, dark hands up to elbows (only up to palms in Northern Plains), and tail that is looped backwards while walking. Occurring in a different geographic zone, it occupies a considerably different kind of forest than Northern Plains, though in terms of behaviour they are fairly similar. **DIET** Leaves, fruits, flowers and buds. **DISTRIBUTION** Western Ghats from Gujarat to Karnataka, encompassing half of Maharashtra and all of Karnataka. East of hills around Bangalore replaced by Tufted Grey Langur (see below). **HABITAT** Dry deciduous forests and edges of moist forests. Also associates with humans.

Tufted Grey Langur ■ *Semnopithecus priam*
F Cercopithecidae O Primata

DESCRIPTION Differentiated from other langur species in being crested – its species name (*priam*) refers to the headgear of Priam, King of Troy. However, not all populations

are crested; in some crest is reduced to stub on middle of head. Fur largely smoky-grey but unlike in its neighbour, South-western Langur (see above), its hands are white with black being restricted to fingers. Similar in behaviour to other langurs. **DIET** Leaves, fruits, flowers and buds. **DISTRIBUTION** Andhra Pradesh, south-eastern Karnataka, northern Tamil Nadu and Kerala. **HABITAT** Dry and mixed decidous forests and scrub, and human habitation.

Terai Langur ▪ *Semnopithecus hector*
F Cercopithecidae **O** Primata

DESCRIPTION Found north of range of Northern Plains Langur (see p. 27). This species is structurally different from the other langurs. Body furry with smoky-grey hair, flat head and white ruff covering face. Dark hands. Similar in behaviour to other langurs, and tail is forwardly looped. **DIET** Leaves, fruits, flowers and buds. **DISTRIBUTION** Terai region of Uttarakhand, Uttar Pradesh, Bihar and West Bengal. **HABITAT** Sal and mixed deciduous forests at foothills of Himalaya.

Himalayan Langur ▪ *Semnopithecus schistaceus*
F Cercopithecidae **O** Primata

DESCRIPTION Most similar in appearance to Terai Langur (see above), but differs mainly in its bulk (which is subjective). Body grey but more mauve as against smoky-grey fur of Terai. Ventrally it is much whiter and its ruff appears more hairy. Does not occur in foothills, but more studies are required to validate its altitudinal distribution. **DIET** Leaves, fruits, flowers and agricultural crops. **DISTRIBUTION** Himalaya from Jammu and Kashmir to Sikkim, predominantly at 1,500–3,000m. **HABITAT** Subtropical and temperate montane forests, and around villages at high altitudes.

Capped Langur ■ *Trachypithecus pileatus*
F Cercopithecidae **O** Primata

DESCRIPTION The most colourful primate in the region. Name derives from its cap of black hair that is 'combed' backwards. Coat variable in colour, and four subspecies are recognized. In some underparts are bright orange in colour, in others they have a yellowish or grey wash. Hairs on cheeks are long, giving rise to creamish-red 'whiskers'. Strictly arboreal. Most populations are shy and take flight as soon as they sense human presence. **DIET** Leaves, fruits, flowers and buds. **DISTRIBUTION** Endangered. Northeast India mainly south of Brahmaputra. Two populations (Manas and Nameri) north of river. **HABITAT** Dense and moist broadleaved forests, bamboo and evergreen forests.

Golden Langur

■ *Trachypithecus geei*
F Cercopithecidae **O** Primata

DESCRIPTION Easily the most striking primate in India. Deep cream-coloured coat that turns to bright golden during breeding season in winter. Face, palm and soles are contrasting black. Face has long, spiky hair that forms whiskers from its cheeks. Tail very long; much longer than head-to-body length. Strictly arboreal. Forms small troops of up to 10 individuals. **DIET** Leaves, fruits, flowers and buds. **DISTRIBUTION** Highly range restricted and endemic; found between Rivers Sankosh and Manas in districts of Kokrajhar, Bongaigaon and Dhubri in Assam. Also occurs in neighbouring Bhutan up to 1,600m. **HABITAT** Subtropical moist deciduous forests and montane forests.

Purple-faced Langur

▪ *Trachypithecus vetulus*
F Cercopithecidae **O** Primata

DESCRIPTION Endemic to neighbouring Sri Lanka. Identified by mostly brown coat and dark face, which appears rather purplish and has very long white 'whiskers'. Tail mostly white in colour. Cap of brown hair. A shy species. **DIET** Leaves, fruits, flowers and buds. **DISTRIBUTION** Wet zone of Sri Lanka. Historically more widespread. **HABITAT** Dense and moist forests often close to human habitation.

Phayre's Leaf Monkey ▪ *Trachypithecus phayrei*

F Cercopithecidae **O** Primata

DESCRIPTION Beautiful forest monkey. The alternative name Spectacled Monkey derives from its unique appearance – black body and bold white borders around eyes. Ventral half of coat off white. Tail longer than body, and short but prominent crest. Shy species that takes flight when disturbed. **DIET** Leaves, fruits, flowers and buds, and bamboo leaves. **DISTRIBUTION** Southern part of North-east India in North Cachar, Hailakandi and Karimganj districts of Assam, Tripura and Mizoram. **HABITAT** Mixed moist deciduous forests and bamboo patches at forest edges.

Asian Elephant ■ *Elephas maximus*
F Elephantidae **O** Proboscidea

DESCRIPTION Head is anterio-posteriorly compressed. Eyes small compared with size of head. Back is convex. Dorsal borders of large ears are folded. Distinctive trunk is used for feeding, smelling and dusting. Tip of trunk has one finger-like proboscis. Adult males can weigh up to 5,000kg, and measure above 3m at shoulder; females weigh about 4,000kg. Body colour varies from grey to greyish-black. **DIET** Both browser and grazer, feeding on grasses, bamboo plants, shrubs, bark of trees and fruits. **DISTRIBUTION** Western Ghats, southern India (south of Mysore), Odisha, Bihar, Himalaya range in Uttar Pradesh, Terai Landscape, West Bengal and Assam. **HABITAT** Variety of vegetation types. Thorn scrub to dry and moist deciduous forests, sal forests, alluvial floodplains, evergreen forests, grassland, marshes and riverine forests.

Asiatic Wild Ass ▪ *Equus hemionus*
F Equidae **O** Perissodactyla

DESCRIPTION Males weigh about 240kg, females around 200kg. Shoulder height is about 110–127cm. Summer coat dark greyish to reddish-grey, which becomes greyish to chestnut during winter. Dark brown band extends from mane, fading midway. Underparts are white, which extends halfway to flank. **DIET** Generalist herbivore, consuming coarse plants and grasses. **DISTRIBUTION** Deserts in north-western India around Dhrangadhra Wild Ass Sanctuary in the Little Rann of Kachchh in Gujarat. **HABITAT** Hot salt deserts below sea level.

Tibetan Wild Ass ■ *Equus kiang*
F Equidae **O** Perissodactyla

DESCRIPTION Body weight about 250–300kg, and measures 140cm at shoulder. Thick, blunt muzzle, and thick neck with short, upright mane. Coat chestnut, becoming dark brown in winter and reddish-brown in summer. Dark brown band extends on dorsal side from mane to end of tail. **DIET** Generalist herbivore, consuming coarse plants and grasses. **DISTRIBUTION** Indo-Tibet borders of Ladakh, Jammu and Kashmir, north Sikkim and Himachal Pradesh. **HABITAT** Broad valleys and riverine tracts, around high-altitude lakes and basins with grasses and sedges.

Indian Rhinoceros ▪ *Rhinoceros unicornis*
F Rhinocerotidae **O** Perissodactyla

DESCRIPTION Huge, with a strong build. Measures about 182cm at shoulder. Grey body skin is in the form of shields with heavy folds. Shoulders, hindquarters and flanks dotted with tubercle-like structures. Single horn above snout. **DIET** Grazer, feeding mostly on grasses. **DISTRIBUTION** Duars in West Bengal, Assam and Terai Arc Landscape. **HABITAT** Floodplains, riverine forests, swamps and grassland.

Indian Chevrotain ■ *Moschiola indica*
F Tragulidae **O** Artiodactyla

DESCRIPTION One of the smallest ruminants, with an unusual three-chambered stomach, shoulder height of 25–30cm and rather short limbs. Length of body around 50–58cm, ending in a short tail 3–5cm long. Coat colour reddish-brown with small white spot coming together as stripes along flanks and three prominent white stripes along throat. Crown, forehead and region between eyes are dark brown or blackish. **DIET** Mainly forages on shrubs and herbs, along with fallen fruits on the forest floor. Depends on fruit-based diet that allows rapid gut passage to meet its high energy requirements.

Fruit species eaten include *Terminalia bellerica, Gmelina arborea, Garuga pinnata, Bridelia retusa, Dillenia pentagyna, Ficus bengalensis, F. glomerata, F. mysorensis* and *Grewia tiliifolia*, which are common in the deciduous forests of South India. **DISTRIBUTION** Western Ghats, in Eastern Ghats to Odisha, and in forests of Central India. **HABITAT** Tropical deciduous, moist evergreen and semi-evergreen forests up to 1,850m elevation. Favours moist forests and riverine patches.

Himalayan Musk Deer ■ *Moschus leucogaster*
F Moschidae **O** Artiodactyla

DESCRIPTION Stocky deer about 50cm at shoulder height, with small head and bounding gait. Hindlegs longer than forelegs by about 5cm. Weighs about 13–15kg. Coat colour reddish, with golden underside, inner surfaces of limbs and mid-line of throat. White stripes on throat and orange eye-ring. Limbs dark brown on upperside, becoming light

towards lower end. **DIET** Primary browser that feeds on leaves of woody plants, forbs, lichen, moss, ferns and grasses. Particularly selects oak *Quercus semecarpifolia*, rhododendron *Rhododendron campanulatum*, montane bamboo *Arundinaria* spp., *Gaultheria nummularioides* and *Rubus* spp. in winter. **DISTRIBUTION** Mountains of North-east India, Nepal, Bhutan, in Himalaya from Himachal Pradesh through Uttarakhand. **HABITAT** Mostly above an altitude of 2,500m. Favours alpine scrub, subalpine oak-fir-maple habitat, and birch-rhododendron forests, as well as bamboo forests below alpine zone (2,600–3,000m).

Indian Muntjac ■ *Muntiacus muntjak*
F Cervidae O Artiodactyla

DESCRIPTION Muntjacs are more commonly known as barking deer, and are small, solitary, forest deer species. They have uniform reddish coats without any conspicuous markings. Underside and inner sides of legs slightly paler. Forelimbs longer than hindlimbs. Males have two black lines running along antler pedicles, extending down face. Males marginally larger than females, weighing about 20–22kg. **DIET** Nibblers, feeding mainly on fruits, buds, seeds and fresh leaves. **DISTRIBUTION** Throughout peninsular India, Terai region, North-east India and lower reaches of Himalaya. There are ten recognized subspecies, found in Sri Lanka, India, Pakistan, Nepal, Bhutan, Bangladesh, South China, Indo-China to Malayan peninsula, Sumatra, Borneo, Java, Bali, Lombok and other Indonesian islands. **HABITAT** Dense forest areas, in hilly and moist regions. Found in thick deciduous and evergreen forests.

Sambar ▪ *Rusa unicolor*
F Cervidae **O** Artiodactyla

DESCRIPTION Summer coat (appears from about May) brown to chestnut-brown. Underside, rump and inner parts of legs light brown. Winter coat (October–May) grey brown, and rutting male is almost black. Tip of tail is black. Conspicuous ruff of hair around neck in both stags and hinds. Fawns have chestnut-brown coats without any spots.

Adults have a shoulder height of 140–150cm. Male weighs about 225–320kg, female around 135–225kg. Only males have antlers, which have three tines on each side and are annually shed. **DIET** Not very specialized in food requirements. Diet mainly consists of browse and grass. In some regions Sambar are primarily grazers, depending on grasses associated with wetlands and grassland. Sometimes they consume fruits like *Balanites aegyptiaca*, *Zizyphus mauritiana* and *Z. xylopyrus*. **DISTRIBUTION** Throughout India, except arid regions of Western India. Found up to treeline in Dodital, Kedarnath Wildlife Sanctuary, Bhutan and Arunachal Pradesh. **HABITAT** Favours dry-thorn, dry deciduous, moist deciduous, semi-evergreen and evergreen forests. Generally found in slightly undulating, hilly areas.

Hangul ▪ *Cervus elaphus hanglu*
F Cervidae **O** Artiodactyla

DESCRIPTION Large brownish deer with black tail. Small, orange-white rump-patch bordered by black band. Underside inner ears are whitish. Antlers usually have five tines. **DIET** Main diet consists of leaves, twigs and grasses. **DISTRIBUTION** Dachigam National Park, Gurez, Waragat-Naranag and Chandaji Nullah in Jammu and Kashmir. **HABITAT** Moist broadleaved, coniferous forests and alpine meadows.

Brow-antlered Deer ■ *Recervus eldii*
F Cervidae **O** Artiodactyla

DESCRIPTION Moderately built deer with shoulder length of 80–90cm in both sexes. Males heavier than females, weighing about 90–130kg, and females weigh around 60–90kg. Summer coat reddish-brown in both sexes, becoming greyish-brown during winter. Prominent pre-orbital scent gland below each eye, and subcaudal scent glands at base of tail in both sexes. **DIET** Mainly grazers and additionally feed on fruits, cultivated crops (rice), forbs and similar plants. **DISTRIBUTION** Restricted to Keibul Lamjao National Park in Manipur, India. Outside India, found in Cambodia, China and Laos. **HABITAT** Dry deciduous forests and open, grassy areas.

Swamp Deer ■ *Recervus duvaucelii*
F Cervidae **O** Artiodactyla

DESCRIPTION Large-bodied deer with shoulder height of 110–125cm. Adults weigh about 170–180kg. Summer coat reddish-brown, with white spots along spine in both sexes. Underside and inner parts of legs are whitish. Winter pelage greyish-brown, and a

lot darker during rut. Ears roundish with tufts of thick hair. As the name suggests, it has a 12-point antler pattern. Weight of antlers and tine numbers vary with age and sometimes with nutrition. **DIET** Feeds mostly on grasses. **DISTRIBUTION** Patchy distribution in Kheri, Pilibhit, Dudhwa National Park and adjacent areas in south-western Nepal, Suklaphanta Sanctuary, in Manas and Kaziranga National Park in Assam with a few individuals in Laokhowa Sanctuary. Also recorded in Kanha National Park in Madhya Pradesh, and Jhilmil Jheel near Haridwar, Uttarakhand. **HABITAT** Lowland swampy grassland with an abundance of moist grassland vegetation.

Hog Deer ■ *Axis porcinus*
F Cervidae **O** Artiodactyla

DESCRIPTION Small, solitary deer considered to be one of the primitive deer species. Looks similar to muntjacs, but is slightly larger. Summer body coat light brown to reddish-brown. Dark dorsal stripe with parallel white spots along both sides, which is not apparent during winter as coat colour changes to darker brown. Upper portions of forelegs are dark in stags. Underside, throat and insides of legs paler than body. Stags are heavily built with three-tined antlers. **DIET** Mainly grazers of herbs and short grasses, and browse on shrubs. **DISTRIBUTION** Eastern subspecies A. *porcinus annamiticus* distributed in Thailand, Laos, Cambodia and Vietnam. Western subspecies A. *porcinus porcinus* found across Pakistan, northern India, Nepal, Bhutan and Myanmar. **HABITAT** Moist, tall grassland and riverine forests. Prefers low-lying areas.

Spotted Deer ■ *Axis axis*
F Cervidae **O** Artiodactyla

DESCRIPTION Distinctive deer with white spots sprinkled on rufous coat. Coat provides good camouflage for the animal in forest undergrowth. Sexual dimorphism exists in this species, with males being darker brown in colour than females and weighing about 70kg, and females weighing around 50kg. Stag has a pair of antlers attached to pedicels. Adult

males have antlers more than 60cm long. **DIET** Primary grazer as well as browser. Favours grass sprouts, blade tips and flowering heads of tall grasses. Also consumes leaves, flowers and fruits mainly during dry season. **DISTRIBUTION** Throughout India at 8–30° N. Occurs in dry forests of Gujarat and Rajasthan. In eastern part, found in Sunderbans, Western Assam. **HABITAT** Widespread in forests and grassland both in flat terrains and on hill slopes. Habitat preference changes with seasons and availability of food.

Gaur ▪ *Bos gaurus*
F Bovidae **O** Artiodactyla

DESCRIPTION One of the heaviest land mammals, weighing about 600–940kg, with a shoulder height of 1.6–1.9m. Muscular ridge on shoulders is one of the most striking morphological features. Both sexes have crescent-shaped horns with sharp, tapering ends. Forelegs and hindlegs white or yellowish to knees – an identifying feature of this species. **DIET** Grazers and browsers, preferring young and mature tree leaves, shrubs, herbs and bamboo shoots. **DISTRIBUTION** Forested tracts of India in isolated pockets in Western Ghats, Central Indian Highlands and North-eastern Himalaya. **HABITAT** Tropical moist deciduous and dry deciduous forests.

Mithun ▪ *Bos frontalis*
F Bovidae **O** Artiodactyla

DESCRIPTION Smaller than Gaur (see p. 48) but with white 'stockings'. Body colour varies from reddish-brown to blackish-brown. Can weigh 650–1,000kg. Both sexes have pairs of horns. Muscular ridge on shoulders. **DIET** Browsers and grazers, feeding on grasses, forbs and leaves. **DISTRIBUTION** North-east India in hills of Tripura, Mizoram, Arunachal Pradesh, Mizoram and Nagaland. **HABITAT** Hill forests and grassy clearings.

Wild Yak ▪ *Bos mutus*
F Bovidae **O** Artiodactyla

DESCRIPTION Dense undercoat covered with long, dark brown outer hair. Entire coat has a shaggy appearance and covers half the legs. Can weigh up to 1,000kg, with shoulder height of more than 2m. **DIET** Grazes on grasses, herbs, mosses and lichens. **DISTRIBUTION** Chang Chenmo Valley of Ladakh (eastern Kashmir). **HABITAT** Treeless uplands, including plains, hills, mountains and alpine meadows up to 5,400m.

Wild Buffalo ■ *Bubalus arnee*
F Bovidae **O** Artiodactyla

DESCRIPTION Bigger and heavier than domestic buffalo. Can weigh up to 700–1200kg, with shoulder height of 150–190cm. Body colour grey to black. Both sexes have horns with wide bases, which spread to about 2m. Forehead covered with tuft of hair. **DIET** Grazers, feeding mainly on grasses and sedges. Also consume fruits and bark, and browse on shrubs. **DISTRIBUTION** Distributed in small pockets in Assam, Arunachal Pradesh, Meghalaya, Chhattisgarh, Madhya Pradesh and Odisha. **HABITAT** Wet grassland, swamps and densely vegetated riversides.

Nilgai ▪ *Boselaphus tragocamelus*
F Bovidae **O** Artiodactyla

DESCRIPTION Adult males steel-grey or blue-grey; females light brown. Dark and white markings on head, underparts and tail in both sexes. Inner side of ear is white. Tuft of hair on ventral side of neck. Adult can weigh up to 200kg, with shoulder height of about 130–140cm. **DIET** Feeds on variable proportions of grass, herbs and browse. Sometimes feeds on fallen leaves, fruits and flowers. **DISTRIBUTION** From foothills of Himalaya through Central India, to southern districts of Andhra Pradesh. Found in 16 protected areas. **HABITAT** Open areas with undulating or flat terrain.

Four-horned Antelope ▪ *Tetracerus quadricornis*
F Bovidae O Artiodactyla

DESCRIPTION The only ungulate with four horns. Front pair shorter than rear pair. Small antelope weighing about 20kg, with shoulder height of 55–60cm. No sexual dimorphism. Coat golden-brown with coarse hair; it becomes darker during monsoon and winter. Neck and underparts paler than rest of body. Rounded ears with whitish tufts towards inner edges. **DIET** Nibblers and browsers, feeding on fruits, flowers, pods and fresh browse. **DISTRIBUTION** Widespread throughout Central and Southern India. **HABITAT** The only forest antelope found in India. Favours deciduous habitat with open areas.

Chinkara ■ *Gazella bennettii*
F Bovidae **O** Artiodactyla

DESCRIPTION Small, graceful gazelle weighing about 23kg, and 65cm high at shoulder. Males have cylindrical, slender black horns with rings, measuring about 25–30cm in length. Females do not have horns, but only stubs. Coat chestnut with white underside and black tail. Dark brown line extends from eyes to mouth. Knees have tuft of hair. **DIET** Mostly browsers, but sometimes feed on grasses and fruits. **DISTRIBUTION** Found to Bihar in north-eastern border and Karnataka-Andhra Pradesh in southern limit. **HABITAT** Open scrubland, plain broken terrain, riverbeds and sandy areas.

Blackbuck ■ *Antilope cervicapra*
F Bovidae **O** Artiodactyla

DESCRIPTION Males strikingly coloured in black and white, with pair of spiralling horns that are about 70cm in length and ringed. Females have brown coat and white underside. White eye-patch in both sexes. Weight about 34–45kg, and shoulder height of 70–75cm. **DIET** Primarily grazers, feeding on fresh grass shoots. In winter they sometimes feed on seed pods of *Prosopis cineraria* and *P. juliflora*. **DISTRIBUTION** A few scattered population in Western through Central to Southern India. **HABITAT** Known to favour open plain areas, but found in semi-arid grassland, open scrub, grassy clearings and open forest.

Tibetan Antelope ▪ *Panthalops hodgsonii*
F Bovidae **O** Artiodactyla

DESCRIPTION Males measure about 80cm at shoulder. Muzzle characteristically swollen in males – this is understood to be a high-altitude adaptation. Antelope have a gland under the eyes. Body covered in thick fawn wool coat, and undersides are white. Dark brown face and stripe down each leg. Lyre-shaped, long horns in males only. **DIET** Grazers of sprouting grass. **DISTRIBUTION** Chang Chen Mo Valley in northern Ladakh in India, which is continued to the great desert in northern Tibet. **HABITAT** Cold desert areas, grassy flats and ravines.

Tibetan Gazelle ■ *Procapra picticaudata*
F Bovidae **O** Artiodactyla

DESCRIPTION Males and females measure 50–60cm at shoulder and weigh 10–15kg. Males have long, tapering horns with ridges, reaching lengths of 30cm. Females have no horns. Body coat greyish-brown, with summer coat being greyer than winter one. Short tail in white rump-patch. Thin, long legs for running in order to escape from predators. **DIET** Feeds on grass, forbs and sedges. **DISTRIBUTION** Sikkim and Ladakh. **HABITAT** High-elevation steppe and alpine regions.

Asiatic Ibex ▪ *Capra sibirica*
F Bovidae **O** Artiodactyla

DESCRIPTION Sturdy, goat-like animal. Males weigh about 80–130kg, while females are around 50–60kg in weight. They stand at about 70–100cm at shoulder height. Face short with longer beard in males than in females. Stocky legs that help them to climb rocks. Sexual dimorphism is present, with winter coat of male being dark brownish with a white saddle. Legs, thighs and underparts also have a whitish colour in some males. Females have lighter coats with little white areas. During summer, their coats become a lot paler. Horns backwardly arched or scimitar shaped, with prominent ridges. **DIET** Feeds on grasses, sedges, forbs and sometimes shrubs. **DISTRIBUTION** Shyok Valley in Ladakh to Sutlej Gorge in Himachal Pradesh. **HABITAT** Rugged mountains of cold, arid regions. Mostly alpine scrub and dry alpine steppe vegetation.

Ladakh Urial ▪ *Ovis orientalis*
F Bovidae **O** Artiodactyla

DESCRIPTION Smaller than wild sheep, measuring about 90cm at shoulder. Summer coat rufous-grey that turns into brownish-grey during winter. Adult ram has a ruff growing from either side of chin and extending down throat region. Horns are curvy and wrinkled. **DIET** Mainly feeds on grasses, leaves and sometimes shrubs. **DISTRIBUTION** Ladakh, eastwards to northern Tibet, Punjab, Sind and Baluchistan. **HABITAT** Steep, grassy hill slopes, open, grassy mountain slopes and rocky, scrub-covered hills.

Bharal ▪ *Pseudois nayaur*
F Bovidae **O** Artiodactyla

DESCRIPTION Stockily built species with strong but short legs to survive in rugged terrain. It has characteristics of both sheep and goats. Males can weigh up to 60–70kg, and measure 80–90cm at shoulder. Females weigh about 50kg. Males have massive horns that

curve outwards and have distinct annual rings, while females have smaller horns with less distinct annual rings. Body coat greyish-buff on dorsal side, white underside and legs, and often a horizontal black stripe on flank. **DIET** Mixed feeder, consuming grasses, herbs and shrubs. **DISTRIBUTION** Throughout Himalaya in Lahaul, Spiti (Himachal Pradesh), Zanskar and Nubra in Ladakh, Hemis National Park (Jammu and Kashmir), Govind Pashu Vihar, Kedarnath Wildlife Sanctuary and Nanda Devi National Park (Uttarakhand), and Sikkim, Western Arunachal Pradesh. **HABITAT** Mountain pastures, steppe vegetation and subalpine slopes, close to cliffs.

Himalayan Tahr ▪ *Hemitragus jemlahicus*
F Bovidae **O** Artiodactyla

DESCRIPTION Heavy-bodied animal with robust limbs for climbing rocky mountains. Adult male measures about 91–102cm at shoulder. Body covered with tangled coarse hair.

Prominent ruff and hair, and rump also covered with long mantle of hair. Body coat deep reddish-brown, and adult males are darker. Insides of legs are white. Females have fewer ruffs and are paler in body colour than males. Both sexes have horns that grow to about 40cm and are curved backwards. **DIET** Primary grazer, feeding on grasses, sedges, herbs, ferns and mosses. Sometimes feeds on montane bamboo and lichens. **DISTRIBUTION** Patchy distribution in southern Greater Himalaya from Pir Panjal range in North India to Nepal and Sikkim. **HABITAT** Mountainous habitat from mid-temperate (2,500–2,700m) to alpine above treeline.

Nilgiri Tahr ▪ *Nilgiritragus hylocrius*
F Bovidae **O** Artiodactyla

DESCRIPTION Shoulder height about 110cm, and weight about 100kg. Adult males have dark brown to black coats. Females are paler grey or tan, which is a good blend with the gneiss cliffs in their habitat. Both sexes have horns, with those of males being heavier and longer than those of females. **DIET** Primary grazer, but also feeds on forbs and shrubs. **DISTRIBUTION** States of Kerala and Tamil Nadu along a narrow stretch in Western Ghats. **HABITAT** Grassland areas close to rock faces and sheer cliffs at 1,200–2,700m.

Goal ■ *Nemorhaedus goral*
F Bovidae **O** Artiodactyla

DESCRIPTION Goat-like species with short, stumpy legs adapted for jumping and climbing. Weighs about 20–30kg, and has a shoulder height of 58–70cm. Both sexes have backwards-curved, pointed horns. Horns in males are thicker at base than those of females, and can be up to 23cm in length. **DIET** Primary grazer, feeding on lichens, grasses, tender stems and leaves. **DISTRIBUTION** Himalayan Goral *N. goral bedfordi* found in north-western India, eastern counterpart *N. g. hodgsoni* in north-eastern India. Chinese Goral *N. caudatus* also found in north-eastern India. **HABITAT** Tropical moist deciduous forests, subtropical pine forests, montane wet evergreen forests up to alpine pastures and birch forests.

Himalayan Serow ■ *Capricornis thar*
F Bovidae **O** Artiodactyla

DESCRIPTION Medium-sized species with large head, short limbs and broad, thick neck. Adult weighs around 90kg and measures almost 100cm at shoulder. Limbs chestnut at the beginning, and whitish below; shoulders and flanks are reddish, and head and mane are black. **DIET** Browser, feeding on shrubs. **DISTRIBUTION** At 200–3,600m in Western Himalaya and North-east India. **HABITAT** Moist, forested gorges and mountain habitats. Hilly evergreen montane forests.

Mishmi Takin ■ *Budorcas taxicolor*
F Bovidae **O** Artiodactyla

DESCRIPTION Large and heavy-bodied animal with very strong, thick legs. Characteristic convex face and stout, broad neck. Males can weigh up to 300kg, and shoulder height is about 140cm. Body coat has a shaggy appearance and is yellowish-grey. Flanks are dark brown or black. Males have more prominent horns than females. **DIET** Generalist browser. Feeds on steep hillside scrub and subalpine patches. **DISTRIBUTION** Sikkim, Arunachal Pradesh (Tawang, Siang Valley and Lower Dibang Valley). **HABITAT** Steep mountain forests above 1,300m. Dense, forested habitats near coniferous forests, broadleaved forests and sometimes evergreen forests.

Pygmy Hog ■ *Porcula salvania*
F Suidae O Artiodactyla

DESCRIPTION Males weigh about
8–10kg, and have a shoulder height of
25cm. Females weigh 6–8kg, and measure
20–22cm at shoulder. Hindlimbs longer
than forelimbs, and short ears. Pelage
blackish-brown along mid-dorsal line. Dark
facial band. Coarse hair behind shoulders.
Underside, inner legs much paler and have
sparse hairs. **DIET** Omnivorous, feeding
on roots and tubers, grass, leaves, shoots,
fruits, seeds, insects and similar items.
DISTRIBUTION Small pockets in foothills
of Himalaya, Manas National Park, Assam-
Bhutan border. **HABITAT** Flat terrain with
sal riverine forests and tall grassland.

Wild Pig ■ *Sus scrofa*
F Suidae **O** Artiodactyla

DESCRIPTION Bulky pig with relatively thin legs. Head adapted for digging, with elongated nose and strong neck muscles. Weighs 60–80kg. Body coat dark greyish with long, coarse hair. Males have mane running down back, which is particularly noticeable during autumn and winter. Canine teeth much more prominent in males than in females, and grow throughout life. Piglets have stripes on back. **DIET** Fruits, seeds, roots and tubers, and wide range of other foods including some animal matter. **DISTRIBUTION** Throughout India. **HABITAT** Deciduous and mixed forests.

Tiger ▪ *Panthera tigris*
F Felidae **O** Carnivora

DESCRIPTION Largest of all extant cats, about 300cm in length and weighing around 250kg as an adult. Coat colour ochre to orangish-yellow. White underside, inner limbs and cheeks. Black stripes all over body, which are used for identifying individuals. Stripes aid Tigers in ambush and camouflage. **DIET** Strict carnivore, mainly feeding on ungulates, cervids and livestock. **DISTRIBUTION** Patchy distribution in South, Central and North-east India, as well as the Terai Arc Landscape and Sunderbans. **HABITAT** Grassland, hill forests, and deciduous and evergreen forests.

A family of tigers playing

Lion ■ *Panthera leo*
F Felidae **O** Carnivora

DESCRIPTION The only cat with a mane and tufted tail. Mane colour and growth vary according to age. Longitudinal belly skin-folds distinguish it from African Lion. Body colour yellow-ochre. Whisker patterns are used to identify individuals. **DIET** Strict carnivore, mainly feeding on ungulates, cervids and livestock. **DISTRIBUTION** Gir forests in Gujarat. **HABITAT** Tropical dry deciduous forests interspersed with thorn forests.

Leopard ▪ *Panthera pardus*
F Felidae **O** Carnivora

DESCRIPTION Weight about a quarter of that of Tigers and Lions. Agile and built for speed, useful for hunting. Coat colour ranges from pale yellow to ochre. White underside of body and tail. Black rosettes all over body help in individual identification. **DIET** Strict carnivore, mainly feeding on ungulates, cervids and livestock. **DISTRIBUTION** Throughout India except arid regions in Western India. Can also be found in human habitations. **HABITAT** Tropical grassland, woodland, and dry deciduous, moist deciduous and evergreen forests. Conifer and broadleaved forests in Himalaya up to 3,000m.

Snow Leopard ▪ *Uncia uncia*
F Felidae **O** Carnivora

DESCRIPTION Medium to small-sized cat. Measures about 60cm at shoulder, and weighs around 40–50kg. Pelage smoky-grey, patterned with dark grey rosettes and spots. Forelimbs are short as an adaptation to rocky mountainous habitat. Thick, long tail acts as a balancer when climbing up mountain cliffs. **DIET** Hunts for small animals like marmots, Bharal, Goral and livestock. **DISTRIBUTION** Himalaya in states of Jammu and Kashmir, Himachal Pradesh, Sikkim, Uttarakhand and Arunachal Pradesh, at 1,800–5,800m. **HABITAT** Subalpine and alpine zones in mountains.

Asian Golden Cat ■ *Catopuma temminckii*
F Felidae **O** Carnivora

DESCRIPTION Largest of the small cats in India, in general appearance resembling a Puma *Puma concolor* (of America) Coat golden-brown and unmarked. Two distinctive moustache-like white stripes on face, and forehead has longitudinal markings. Melanistic black forms are also known, which can be identified by distinctive white moustachial stripes. Tail long and not bushy. Nocturnal, and chiefly terrestrial but can also climb well. Not much known about behaviour. **DIET** Small deer, rodents and birds; also attacks domestic sheep, poultry and similar. Can hunt animals larger than itself. **DISTRIBUTION** Uncommon. North Bengal, Sikkim and North-east India. **HABITAT** Tropical and subtropical evergreen and moist deciduous forests up to 3,000m

Marbled Cat ■ *Pardofelis marmorata*
F Felidae **O** Carnivora

DESCRIPTION Miniature version of Clouded Leopard (see opposite), with long, furry tail. Coat greyish-brown or reddish, replete with blotches and irregular rosettes that give it a marbled appearance. Blotches have pale borders, unlike black borders in Clouded Leopard. Face has stripes running from eyes through cheeks, and longitudinal stripes on forehead. Legs strongly spotted. Nocturnal and chiefly arboreal. Habits largely unknown. **DIET** Supposedly mainly birds (due to its arboreal nature), but also takes rodents and small mammals. **DISTRIBUTION** Rare. North Bengal, Sikkim and North-east India. **HABITAT** Tropical and subtropical evergreen and moist deciduous forests.

Clouded Leopard ■ *Neofelis nebulosa*
F Felidae **O** Carnivora

DESCRIPTION Extremely elegant cat with warm ochre coat with cloud-like grey markings. 'Clouds' turn into oval black spots along hindquarters, reducing further to black spots in very long tail. Face has usual stripes on cheeks, and two bold, longitudinal stripes on forehead. Nose prominently pink. The largest canine teeth among all living cats. Nocturnal and very secretive. Chiefly arboreal, and probably raises its kittens in tree hollows. **DIET** Probably monkeys, lorises and flying squirrels; also birds and similar. **DISTRIBUTION** North Bengal, Sikkim and North-east India. **HABITAT** Tropical and subtropical evergreen and moist deciduous forests; also reported from dry deciduous forests.

Caracal ■ *Caracal caracal*
F Felidae **O** Carnivora

DESCRIPTION Tall and sleek cat. In general build fairly similar to Cheetah *Acinonyx jubatus* (of Africa), and also known for its agility and quick pace. Coat tawny or rufous in colour. Fur short, thick and soft. Distinctive feature of the Caracal is long tuft of hair at tip of ear, which is black on the back. Tail not very long and tall legs unmarked. Chiefly nocturnal; probably spends the day hiding among scrub cover or in burrows. Not much known of its habits. **DIET** Birds, hares, rodents and reptiles; can also hunt small deer or gazelle. **DISTRIBUTION** Rare. Arid north-west extending eastwards into Madhya Pradesh. **HABITAT** Arid and semi-arid scrub forests and open rocky areas.

Lynx ■ *Lynx lynx*
F Felidae **O** Carnivora

DESCRIPTION Graceful cat with ear-tufts, which are generally not as long as those of Caracal (see p. 83). Tail short and stubby. Winter coat thick, woolly and sandy-grey; summer coat light and tawny. Shelters in willow scrub, reeds and tall grass, and hunts by stealth. Probably active during both the day and at night. **DIET** Anything that it can overcome: hares, marmots and partridges, as well as domestic goats and sheep. **DISTRIBUTION** Rare. Trans-Himalaya of Jammu and Kashmir, and Sikkim. **HABITAT** Scrub and rocky outcrops in cold deserts.

Pallas's Cat ■ *Octocolobus manul*
F Felidae **O** Carnivora

DESCRIPTION A peculiar cat, with a distinctive broad head, low forehead, short, widely separated ears, furry mane and long whiskers. Eyes have white borders, giving it a spectacled appearance. Coat grey with thick, long fur. Probably shelters among rocks or in burrows of foxes and marmots. Very secretive and little is known of its behaviour in the wild. **DIET** Rodents and birds. **DISTRIBUTION** Rare. Trans-Himalaya of Jammu and Kashmir, and Sikkim. **HABITAT** Scrub and rocky outcrops in cold deserts.

Jungle Cat ■ *Felis chaus*
F Felidae **O** Carnivora

DESCRIPTION The most common wild cat in India. Distinctive appearance with rather long legs making it appear tall, and relatively short tail. Coat generally uniform sandy-grey to fawn in colour. Legs faintly banded with black; tail also ringed, ending in black tip. Kittens marked all over body, and easily mistaken for domestic cat kittens. Short tuft of hair at tip of ear. Highly versatile and adaptable. Largely nocturnal, but may be active during the day in places with relatively little human disturbance. Rests among dense cover or in burrows. **DIET** Versatile and eats anything it can procure – lizards, mice, birds and frogs, and also known to hunt porcupines. **DISTRIBUTION** Common. Widespread throughout mainland India. **HABITAT** Wide variety of habitats – arid and semi-arid scrub forests, dry and moist deciduous forests, and montane evergreen forests up to 3,000m.

Desert Cat ▪ *Felis sylvestris*
F Felidae **O** Carnivora

DESCRIPTION Small cat, much like domestic cat in size and appearance. Fur pale yellowish infused with grey, and with black spots throughout. Terminal half of tail ringed with black, and legs also have black stripes. Horizontal stripes on cheeks and longitudinal stripes on forehead. Lives in network of burrows during the day. Chiefly nocturnal and not well studied. **DIET** Gerbils and jirds, and birds and reptiles. **DISTRIBUTION** Uncommon. Arid parts of north-west, particularly in Rann of Kutch in Gujarat, and Desert National Park in Rajasthan. **HABITAT** Scrub forests, deserts and dunes, and cultivated tracts.

Fishing Cat ▪ *Prionailurus viverrinus*
F Felidae **O** Carnivora

DESCRIPTION Large and robust cat. Fur olive-grey with spots neatly arranged in rows. Tail short, stubby and ringed. Stripes on cheeks and longitudinal stripes on forehead. Short ears with white spot on back. Partly webbed feet. As its name suggests, largely aquatic and hunts for fish in shallow waters, or from reed beds and grassy swamps. Nocturnal. **DIET** Primarily fish and molluscs, but also hunts rodents and even domestic goats. **DISTRIBUTION** Rare. Terai floodplains, North-east India, mangroves and wetlands in South Bengal and Andhra Pradesh, and Southern Western Ghats. Ranges disjunct. **HABITAT** Undisturbed wetlands with tall grass, mangroves and freshwater pools in dense forests.

Leopard Cat ■ *Prionailurus bengalensis*
F Felidae **O** Carnivora

DESCRIPTION Small, forest-dwelling cat that looks like a miniature Leopard (see p. 78). Fur yellowish and strongly marked with black spots and blotches. Tail has combination of rings and spots, and legs are spotted. Face has typical horizontal stripes on cheeks and longitudinal stripes on forehead. White spot on back of ear. Nocturnal and semi-arboreal. Known to live in tree hollows during the day. **DIET** Birds, rodents, frogs and lizards; also known to take poultry when close to human habitation. **DISTRIBUTION** Throughout Himalaya (up to 3,000m), North-east India and Southern Western Ghats. Scattered records from rest of peninsula indicate that it may be more widespread. **HABITAT** Moist deciduous and evergreen forests, grassland and mangroves.

Rusty-spotted Cat ■ *Prionailurus rubiginosus*
F Felidae **O** Carnivora

DESCRIPTION Smallest wild cat in the world, half the size of a domestic cat and easily confused with kitten of a Jungle Cat (see p. 85). Fawn coat with rusty-brown spots (which often become faint and diffuse) arranged neatly. Eyes faintly ringed with white. Two longitudinal stripes on forehead. Nocturnal and fairly tolerant of human presence, yet an uncommon species. **DIET** Mainly small birds and rodents. **DISTRIBUTION** Peninsular India up to Rajasthan and Madhya Pradesh in north. **HABITAT** Dry scrub, rocky outcrops and dry deciduous forests, often near human habitation; avoids dense and wet forests.

Himalayan Palm Civet
▪ *Paguma larvata*
F Viverridae **O** Carnivora

DESCRIPTION No spots on body and distinctive white whiskers. Body coat grey or tawny, and white underside. Markings on face vary, with a white mark on forehead and below ears. Round black blotch under eye. Thick, unmarked black tail, and black chin and throat. **DIET** Fruits, vegetables, small birds, rodents and similar. **DISTRIBUTION** Assam, Jammu, Kashmir, and Central and Eastern Himalaya. **HABITAT** Hill forests and mountains up to 2,500m.

Common Palm Civet ▪ *Paradoxurus hermaphroditus*
F Viverridae **O** Carnivora

DESCRIPTION Body colour black or brownish-black. Broad faint stripes on either side of body. Black limbs and variable facial markings. Whitish patch below eye and either side of nose. **DIET** Mostly fruits, and sometimes vegetables, small birds, rodents and similar. **DISTRIBUTION** Throughout India, except Himalaya and arid region. **HABITAT** Deciduous, evergreen and scrub forests. Found around human habitations.

Brown Palm Civet

▪ *Paradoxurus jerdoni*
F Viverridae **O** Carnivora

DESCRIPTION Very similar to Common Palm Civet (see opposite), but without any facial markings. Coat uniform chocolate-brown in colour, and grey flanks. Tail, limbs and head region dark blackish. **DIET** Mostly fruits, and sometimes vegetables, small birds, rodents and similar. **DISTRIBUTION** Western Ghats in Goa, Tamil Nadu, Nilgiris and Anamalais. **HABITAT** Wet evergreen forests.

Small Indian Civet ▪ *Viverricula indica*

F Viverridae **O** Carnivora

DESCRIPTION Base body colour greyish-brown, with black spots arranged in rows in flank region. Black lines and streaks on back. Tail has black rings. Underside whitish up to throat region. **DIET** Mostly fruits, and sometimes vegetables, small birds, rodents and similar. **DISTRIBUTION** Throughout India except high mountains. **HABITAT** Grass or scrubland, semi-evergreen, deciduous and bamboo forests, and riverine areas.

Large Indian Civet

▪ *Viverra zibetha*
F Viverridae **O** Carnivora

DESCRIPTION Long, compressed body, and short legs. Distinctive black crest of long hair runs along back. Base body colour greyish-brown, with black bands and rosettes. Black-ringed tail and dark limbs. **DIET** Small birds, rodents, insects and sometimes fruits. **DISTRIBUTION** Sikkim, upper Bengal and Assam. **HABITAT** Hills, moist deciduous and evergreen forests, and scrubland.

Binturong ▪ *Actitis binturong*
F Viverridae **O** Carnivora

DESCRIPTION Bear-like civet species. Shaggy body coat, and tail thicker than body. Grizzled black-and-white coat, and white whiskers. **DIET** Small birds, rodents, insects and sometimes fruits. **DISTRIBUTION** North-east India. **HABITAT** Arboreal, living in dense forests.

Grey Mongoose ■ *Herpestes edwardsii*
F Herpestidae **O** Carnivora

DESCRIPTION Tawny to grey-coloured mongoose. Body densely furred with a salt-pepper appearance. Tail long and with white or pale tip. Ruddy face and limbs. Individuals from

north and arid north-west paler in colour than darker ones from south. Shy species, but a bold and fierce hunter. Generally seen individually or in groups of 3–4 comprising mother and pups. Lives in hedgerows, thickets, tree holes, burrows and termite mounds, and is mostly diurnal. **DIET** Omnivorous, eating anything that it can procure. Efficient hunter of rodents, snakes, frogs, lizards and similar. **DISTRIBUTION** Throughout India, except High Himalaya, and Andaman and Nicobar Islands. **HABITAT** Cosmopolitan; open scrub and bushes, jungles, cultivation and around human habitation. Avoids dense forests.

Ruddy Mongoose ■ *Herpestes smithii*
F Herpestidae **O** Carnivora

DESCRIPTION Similar to Grey Mongoose (see above). Tawny to grey in colour but generally more fawn coloured than Grey Mongoose. Underparts have reddish-brown

infusions. Easily distinguished by tail, which has a bushy black tip curved upwards. Legs dark rufous tending to black. Shy; generally solitary or in groups of one mother and 2–3 young. Lives in hedgerows, thickets, tree holes, burrows and termite mounds, and is mostly diurnal. **DIET** Omnivorous, but mainly feeds on small mammals, reptiles, birds, insects and arachnids. **DISTRIBUTION** Peninsular India, going north up to Delhi and Haryana. **HABITAT** Dry deciduous and partly moist forests; avoids human habitation.

Small Indian Mongoose ■ *Herpestes auropunctatus*
F Herpestidae **O** Carnivora

DESCRIPTION Nearly half the size of Grey Mongoose (see p. 91), with a golden-tawny coat. Muzzle black with a pink nose. Tail relatively short and tip is either dark or concolourous to rest of tail. Very active and moves around swiftly. Generally found solitarily or in group of mother with her pups. Lives within bushes and burrows. **DIET** Highly varied; omnivorous but feeds mostly on small mammals, reptiles, birds, insects and arachnids. **DISTRIBUTION** Northern plains south to Gujarat and northern Madhya Pradesh and plains of North-east India. **HABITAT** Highly adaptable; lives in open scrub and hedges, desert and human habitation.

Marsh Mongoose ■ *Herpestes palustris*
F Herpestidae **O** Carnivora

DESCRIPTION Considered by some authorities as a subspecies of Small Indian Mongoose (see above), and similar in appearance to it. Lives in burrows dug by itself and also by other animals. Chiefly diurnal. Mostly solitary, but occasionally becomes gregarious. **DIET** Probably feeds mainly on small crustaceans, molluscs and amphibians; also tubers, and fruits including berries. Hunts by wading in shallow water; cracks shells of molluscs and eggs of birds by throwing them on hard surfaces. **DISTRIBUTION** Endangered and highly restricted; currently known only from East Kolkata Wetlands and Hooghly in West Bengal. An endemic species. **HABITAT** Wetlands with tall grass and reed beds.

Crab-eating Mongoose ■ *Herpestes urva*
F Herpestidae O Carnivora

DESCRIPTION Large and stocky mongoose with a bushy tail. Coat light grey-brown with broad white stripe on back and white stripes on cheeks and neck. Feet black and partly webbed. Partly aquatic, adeptly hunting for crabs, snails, fish and frogs. Cracks open hard shells of crustaceans and molluscs by throwing them at stones. Largely crepuscular and nocturnal.
DIET Crustaceans, molluscs, fish and amphibians.
DISTRIBUTION From Duars of North Bengal to plains of North-east India (rarely in hills). Rare.
HABITAT Streams, shaded rivers and tea gardens.

Stripe-necked Mongoose ■ *Herpestes vitticollis*
F Herpestidae O Carnivora

DESCRIPTION Largest mongoose in India. Large and stocky; fur varies from grey (Northern Western Ghats) to ruddy (Southern Western Ghats). Characteristic black stripes on neck. Head rather greyish; tail bushy, black and slightly upturned. Diurnal; seen scampering along the forest floor and on trails in pursuit of prey.
DIET Opportunistic, eating anything that it can procure, including rodents, crabs and even mammals as large as a mouse-deer.
DISTRIBUTION Western Ghats south of Goa.
HABITAT Dry and moist evergreen forests, plantations and swampy areas; found in partly hilly country.

Striped Hyena ■ *Hyaena hyaena*
F Hyaenidae **O** Carnivora

DESCRIPTION Larger than members of dog family. Shaggy animal with dog-like build, and shorter and weaker hindlegs than forelegs. Muzzle broad, and ears triangular and erect. Crest of long hair all along back, ending in bushy tail; crest is erected when the animal is alert. Grey fur with uneven stripes. Clumsy gait due to the weak hindquarters and sloping body. Vocal, chattering conspicuously in undisturbed areas and during certain seasons. Raises pups in burrows or dens. **DIET** Primarily adapted to feeding on carrion, but can also hunt domestic goats and dogs. **DISTRIBUTION** Throughout dry tracts of peninsular India. **HABITAT** Semi-arid scrub, dry grassland, rocky outcrops and open forests. Not found in dense forests.

Indian Wolf

Tibetan Wolf

Indian Wolf ■ *Canis lupus*
F Canidae **O** Carnivora

DESCRIPTION Largest among canids in India. Indian Wolf (subspecies *pallipes* of Grey Wolf of Europe and America) is tall and sleek, resembling a slim German Shepherd Dog. Long legs, sandy-brown coat of short hair, and thin, long tail with black tip. Tibetan Wolf *C. l. chanco* has a heavier build, and a dense, woolly coat that varies in colour (seasonally) from fawn to grey, and a bushy tail. Lives in small packs of up to 10 individuals, and shelters pups in burrows. Packs also known to follow pastoral nomadic tribes to hunt injured or diseased goats and sheep. **DIET** Gazelle, antelope, hares, domestic goats and sheep. **DISTRIBUTION** Throughout peninsular India in dry regions; Tibetan Wolf occurs in Ladakh. **HABITAT** Dry, open country and grassland; also occasionally dry forests.

Jackal ■ *Canis aureus*
F Canidae **O** Carnivora

DESCRIPTION Meek-looking animal of medium size (smaller than average domestic dog), with short limbs. Coat buffy in colour, heavily grizzled with grey and black hair. Can be distinguished from a wolf by small size, shorter ears and bushy tail. Face also more pointed and acute than long face of a wolf. Larger than Indian Fox (see p. 96). Largely crepuscular and nocturnal; roams the countryside using a quick and 'sly' gait. Vocal, giving out a long-drawn howl, often joined in a playful chorus by nearby compatriots. **DIET** Rodents, birds and other small game, and carrion; particularly fond of *Zizyphus* berries in winter. **DISTRIBUTION** Throughout mainland India. **HABITAT** Adapted to a wide range of habitats, from dry to semi-arid grassland and scrub forests, to dry and moist deciduous forests; avoids dense forests.

Asiatic Wild Dog (Dhole) ■ *Cuon alpinus*
F Canidae **O** Carnivora

DESCRIPTION Nearly the size of a wolf, this elegant, endangered canid is a forest dweller. Handsome brick-red fur with varying amounts of white. In Central Indian individuals, there is extensive white on throat and belly. Individuals from Western Ghats and North-east India are largely entirely rufous. Ears moderate in size, with white hair emanating from them. Does not bark or howl, communicating hunting activity using infrasonic whistles. One of the most sociable animals, almost always seen in packs, which can grow to 20–30 individuals. Vicious hunter that often begins to eat its prey even before it is fully brought down. **DIET** Deer and other large game. **DISTRIBUTION** Discontinuous range; Central India up to Odisha, Western Ghats, parts of Eastern Ghats and North-east India and High Himalaya. Absent from Terai region. **HABITAT** Dry and moist forests in peninsula, and temperate subalpine forests in Himalaya.

Tibetan Sand Fox ■ *Vulpes ferrilata*
F Canidae **O** Carnivora

DESCRIPTION Odd-looking fox with a fluffy face, flat head and furry mane. Short stature, thick, furry coat and large, bushy tail. Coat colour varies from grey or brown in winter, to rufous with iron-coloured flanks in summer. Tail-tip conspicuous white (prominent in summer coat). Lives in burrows. Not a well-studied species, and much remains to be known about its behaviour. **DIET** Pikas and rodents that occur in alpine meadows. **DISTRIBUTION** Trans-Himalaya of Ladakh and Sikkim. **HABITAT** Alpine pastures, and cold desert steppes above treeline.

Indian Fox ■ *Vulpes bengalensis*
F Canidae **O** Carnivora

DESCRIPTION Meek creature, asymmetrically built with slim body and long, bushy tail. Coat generally greyish with extensive grizzling of black and grey. Legs short and slim.

Individuals in arid north-west often have rufous-coloured legs. Ears large, and prominent smudge of black on muzzle. Spends the day in burrows and active mainly at night, though may be active during the day in undisturbed areas. Pups mainly raised in burrows. **DIET** Mainly rodents, birds and insects; also actively feeds on *Zizyphus* berries and melons. **DISTRIBUTION** Throughout peninsular India in dry regions. **HABITAT** Dry, open grassland and semi-arid scrub, often near human habitation.

Red Fox ■ *Vulpes vulpes*
F Canidae **O** Carnivora

DESCRIPTION Beautiful fox, richly
coloured with silky, red or orange fur.
Thick coat of Red Fox, subspecies *V.
v. vulpes*, gives it a rather heavy build,
unlike that of its desert cousin, Desert
Fox, subspecies *V. v. pusilla*, which is thin
and sleek. Tail-tip prominently white in
both subspecies. Large ears. In Himalaya,
Red Fox lives among bushes in coniferous
forests or rocks above treeline. Desert
Fox lives in network of interconnected
burrows. **DIET** Rodents, birds, lizards,
carrion and insects, as well as small
berries. **DISTRIBUTION** Red Fox occurs
throughout Himalaya above 2,000m;
Desert Fox occurs in arid desert and semi-
desert in Gujarat and Rajasthan.

V. v. vulpes

V. v. pusilla

Asiatic Black Bear ■ *Ursus thibetanus*
F Ursidae **O** Carnivora

DESCRIPTION Large, forest-dwelling bear with short, smooth coat and less clumsy gait
than other large bears. Bold white crescent on breast, earning it the name 'Moon Bear'.
Lacks shaggy appearance of Sloth Bear (see p. 98), and has much shorter muzzle. Claws
are black (vs pale claws of Sloth
Bear). Nocturnal; spends its day in
dens and searches for food at night.
May climb trees in search of food.
Either migrates to lower altitudes
or hibernates during winter in
High Himalaya. **DIET** Broad diet,
including insects, grubs, honey, nuts,
plums and apricots, and also carrion
and human refuse from around
army camps. **DISTRIBUTION**
Throughout Himalaya and North-
east India. **HABITAT** Subalpine
meadows, forested hill slopes,
broadleaved and coniferous forests,
and also moist deciduous forests in
North-east India.

Brown Bear ■ *Ursus arctos*
F Ursidae **O** Carnivora

DESCRIPTION Very large bear and the world's largest terrestrial carnivore (Grizzly and Kodiak Bears of North America are subspecies of Brown Bear). Coat thick, and reddish-brown with a silvery tinge (mainly during winter), which turns to rich wood-brown in summer. Prominent hump, and white claws. Due to its size (and also its habitat), it is

terrestrial and spends a lot of time turning over stones in search of food. Hibernates in winter. **DIET** Wide variety of foods, including fresh alpine grass, insects and grubs, voles, marmots, mulberries, walnuts, peaches and apples; occasionally livestock. **DISTRIBUTION** Western Himalaya from Jammu and Kashmir to Uttarakhand (above 3,000m). **HABITAT** Alpine meadows above treeline; avoids forests.

Sloth Bear ■ *Melursus ursinus*
F Ursidae **O** Carnivora

DESCRIPTION Widespread and familiar species with shaggy coat of unkempt, long black hair. Long and broad muzzle that is tan in colour. Claws conspicuously white and long, as an adaptation to enable it to break apart termite mounds, termites being one of its

favourite foods. Most individuals also have a V-shaped white crescent on the chest. Poor vision, and relies mainly on its sense of smell to find its way. Clumsy gait – walks about heavily and with a sense of urgency. An adept climber, it can knock beehives out of trees. Nocturnal, but may be active during the day where there is not too much human disturbance. **DIET** Wide variety of foods, including insects and grubs, and partial to termites; also honey and grubs of honeybee, and fallen flowers of mohwa (*Madhuca latifolia*). **DISTRIBUTION** Peninsular India, parts of North-east India and foothills of Himalaya. **HABITAT** Dry deciduous forests, scrubland and grassland, and rocky outcrops.

Sun Bear ▪ *Helarctos malayanus*
F Ursidae **O** Carnivora

DESCRIPTION Small forest bear – rarest of the bears in India – superficially resembling a miniature Asiatic Black Bear (see p. 97). Sun Bear also has a short and glossy black coat, but differs in having a golden-yellow, U-shaped mark on its chest. Rather short muzzle and pale mask delineating eyes and muzzle. With its small size, bowed front legs and paws turned inwards, it is chiefly arboreal. Makes rough nests or drays to sleep on. Exclusively nocturnal. **DIET** Insects and fruits. **DISTRIBUTION** North-east India, south of Brahmaputra. **HABITAT** Dense and moist broadleaved forests.

Red Panda ▪ *Ailurus fulgens*
F Ailuridae **O** Carnivora

DESCRIPTION One of the most striking and elegant animals found in India. Racoon-like (some taxonomists do place it in the racoon family), with a rounded head, triangular, erect ears and white markings around nose, eyes and ears on chestnut face. Body also chestnut and tail ringed with white. Generally solitary and primarily nocturnal. Chiefly arboreal but often descends to the ground in search of food. Spends most of the day sleeping in tall trees, wrapping itself in its long, bushy tail. **DIET** Eggs, insects, fruits and succulent grasses. **DISTRIBUTION** North Bengal, Sikkim and Eastern Himalaya in Arunachal Pradesh. **HABITAT** Subtropical and temperate broadleaved forests.

Hog Badger ■ *Arctonyx collaris*
F Mustelidae **O** Carnivora

DESCRIPTION A peculiar animal; it is squat and bear-like in habits and gait, and has the body of a pig, but the face and claws of a badger. Snout like that of a pig. Coat thick and dense, woolly, and grey or tawny-grey in colour grizzled with black. Face has bold white markings running up nose and down cheeks. Rare and elusive, with very little known about its behaviour. **DIET** Omnivorous, hunting rodents, birds and small mammals; also eats fruits and tubers. **DISTRIBUTION** North-east India. **HABITAT** Dense and moist deciduous and evergreen forests.

Honey Badger ■ *Mellivora capensis*
F Mustelidae **O** Carnivora

DESCRIPTION Odd-looking creature with an unmistakable and unique appearance. Bear-like in build and habits, and badger-like in ferocity and nature. Colour is unusual – silvery-grey on dorsal half of body, and black on ventral half. Tail short and stubby, and same in colour as body. Musk glands through which it sprays a smelly fluid in defence (much like a skunk). Reputation for being among the boldest animals on Earth, and will not hesitate to

attack anything in self-defence. Though widespread it is shy, elusive and rarely seen. **DIET** Anything that it can procure, including small mammals, birds, reptiles and termites; also eats fruits and tubers. Believed to form associations with honeyguide birds in Africa to hunt larvae of honeybees (though the validity of this claim is challenged by some). This does not occur in India. **DISTRIBUTION** Peninsular India in dry regions. **HABITAT** Dry deciduous forests with little human disturbance, riverine tracts and rocky outcrops.

Yellow-throated Marten ■ *Martes flavigula*
F Mustelidae O Carnivora

DESCRIPTION Body length about 50–60cm, and weight around 2–3kg. Elongated body with short legs. Body coat variegated brown, black and yellow, with dark bands along nape. Throat region yellow with dark bands. **DIET** Birds, small animals and insects. **DISTRIBUTION** Himalaya and Assam hill ranges. **HABITAT** Temperate forests at 1,200–3,000m.

Nilgiri Marten
■ *Martes gwatkinsi*
F Mustelidae O Carnivora

DESCRIPTION Body length about 50–60cm, and weight around 2–3kg. Elongated body with short legs suited to its arboreal nature. Entire body brown to dark brown, with characteristic yellow patch on throat. **DIET** Birds, small mammals and insects. **DISTRIBUTION** Nilgiri Hills and parts of Western Ghats. **HABITAT** Moist tropical rainforests, and moist deciduous and montane evergreen forests away from human habitation.

Smooth-coated Otter ■ *Lutrogale perspicillata*
F Mustelidae O Carnivora

DESCRIPTION Largest of the south Asian otters. Weighs around 7kg. Body coat can be blackish-brown, light brown, rufous or tawny-brown. Underside paler than dorsal side. Whitish chin and throat regions. Webbed and large front and hind paws. Tail end flat with tapering end. **DIET** Fish eater, but also feeds on shrimps, crayfish, crabs, frogs, insects, mudskippers, and birds and rats. **DISTRIBUTION** Throughout India. **HABITAT** Mostly plains. Rocky stretches along rivers, swamp forests, mangroves, estuaries and even semi-arid regions.

Small-clawed Otter ■ *Aonyx cinerea*
F Mustelidae O Carnivora

DESCRIPTION Smallest of the otter species. Distinguished from the other species by narrow and shallow feet, minute claws and emarginated webs. Weighs 3–6kg. Typical dark brown body coat. Ventral side paler than dorsal side, with greyish tinge. Grey or nearly white chin, cheeks, upper lip, sides of neck and throat. **DIET** Feeds on invertebrates like crabs, molluscs, insects and small fish. Sometimes takes rodents, snakes and amphibians. **DISTRIBUTION** Kerala coast, east and north-east region. **HABITAT** Hill streams, mangroves, freshwater swamps, meandering rivers and tidal pools.

Eurasian Otter ■ *Lutra lutra*
F Mustelidae O Carnivora

DESCRIPTION Similar in size to Small-clawed Otter (see opposite). Body length about 90–120cm. Body coat dusky-brown to rust, and underside whitish or greyish. Throat has yellow or sometimes whitish spots. Tail tapered at end and flattened. All four limbs have webs between toes, and strong claws. **DIET** Fish eater, but also feeds on shrimps, crayfish, crabs, frogs, insects, mudskippers, and birds and rats. **DISTRIBUTION** Eastern and southern coasts of India, north-east region and Terai Landscape. **HABITAT** Highland or lowland lakes, rivers, streams, marshes, swamp forests and coastal areas.

Chinese Ferret-badger ■ *Melogale moschata*
F Mustelidae O Carnivora

DESCRIPTION Smallest of all badger species. Can weigh up to 30kg, and average body size is 30–40cm, with tail of 15–20cm. Distinctive facial markings on black face with white forehead. Forelimbs elongated, with claws. Body coat chocolate-brown or greyish-brown, and whitish underside. Prominent stripe along back. **DIET** Feeds on fruits, insects, small animals, amphibians and worms. **DISTRIBUTION** North-east India. **HABITAT** Tropical and subtropical forests and sometimes grassland.

Siberian Weasel ■ *Mustela sibirica*
F Mustelidae O Carnivora

DESCRIPTION Largest weasel in India, identified by uniform orange-brown coat. Face has varying amount of white flecks and patches, and brown near forehead. Chest has a pale patch. Juveniles generally more brown than adults. Very active and constantly on the move. Lives in holes and burrows. An efficient hunter capable of bringing down prey larger than itself. **DIET** Anything that it can procure; mostly small mammals and birds. **DISTRIBUTION** Throughout High Himalaya at 1,500–5,000m. **HABITAT** Dry, sandy valleys above treeline; also in dense forests at lower elevations.

Mountain Weasel ■ *Mustela altaica*
F Mustelidae O Carnivora

DESCRIPTION Probably the most common weasel in India. Chiefly brownish above and creamy-yellow below; greyish above instead of brown in summer. Lips and cheeks have white patches; legs also conspicuously white. Tail uniformly coloured and concolourous to dorsal pelage. Very active and constantly on the move. Lives in holes in walls and burrows in the ground. An efficient hunter capable of killing prey much larger than itself. Partly diurnal and also active at night. **DIET** Varied; mostly small mammals and birds. **DISTRIBUTION** Throughout High Himalaya at 2,000–4,000m. **HABITAT** Dry, sandy valleys above treeline.

Yellow-bellied Weasel ▪ *Mustela kathiah*
F Mustelidae O Carnivora

DESCRIPTION A little-known weasel. Uniform chocolate-brown above from snout to tail, and sulphur-yellow below. Feet generally concolourous to upperparts. Largely nocturnal. An efficient killer like the rest of its tribe. **DIET** Same as that of other weasels. **DISTRIBUTION** Himalaya at 3,000–5,000m, and low-lying areas in North-east India. **HABITAT** Temperate forests and dense evergreen forests; also dry valleys above treeline.

Stoat (Ermine) ▪ *Mustela erminea*
F Mustelidae O Carnivora

DESCRIPTION Smallest weasel in India; chestnut-brown, chocolate-brown or olive above, and bright white below. Moults before winter to pure white plumage. Tail short and has diagnostic brown or black bushy tip that distinguishes it from all other weasels. Active by day and hunts efficiently. **DIET** Same as that of other weasels. **DISTRIBUTION** High altitudes (3,000–4,000m) in Western Himalaya and Sikkim. **HABITAT** Alpine and temperate forests and open, rock-strewn rivers and valleys above treeline.

Spotted Linsang ■ *Prionodon pardicolor*
F Prionodontidae **O** Carnivora

DESCRIPTION Smallest viverrid in India; low-slung yellow body, with uniform black blotches all over; tail has black rings. Dark nuchal stripe on head. Lives in tree hollows. Solitary, nocturnal, shy and rarely seen. **DIET** Small mammals, reptiles and similar. **DISTRIBUTION** North-east India. **HABITAT** Low-lying, dense evergreen rainforests; also found in secondary forests up to 2,700m.

Indian Hare ■ *Lepus nigricollis*
F Leporidae **O** Lagomorpha

DESCRIPTION Common, medium-sized grey to brown hare with large, ovate ears. Upperparts variable in colour depending on geography: pale rufous or grey in northern areas (where it is known as a different subspecies called Rufous-tailed Hare), pale yellowish or sandy-grey in desert areas (known as Desert Hare), and with dark brown or black patch on nape and shoulder in south (known as Black-naped Hare). Roosts in burrows during day and active at night. A fast runner. **DIET** Herbivorous, feeding on dry grass and seeds. **DISTRIBUTION** Throughout mainland India except High Himalaya and dense forests of north-east. **HABITAT** Open scrub, grassy patches and dry forests.

Woolly Hare
■ *Lepus oiostolus*
F Leporidae **O** Lagomorpha

DESCRIPTION Plump and stocky hare with thick, woolly fur. Coat curly, pale overall and grizzled dark brown or black above – ideal for concealment in its snowy or rocky habitat. Ears long with black patches at tips. White eye-ring. Roosts in burrows of marmots. **DIET** High-altitude herbs and grass. **DISTRIBUTION** Trans-Himalaya of Jammu, and Kashmir and Sikkim (above 3,000m). **HABITAT** Alpine meadows and open, rocky terrain.

Hispid Hare
■ *Caprolagus hispidus*
F Leporidae **O** Lagomorpha

DESCRIPTION Stout hare with small hind limbs and hunched posture. Ears very short and emerge just a few inches above body. Fur grizzled with dark hair, and white belly. Moves about slowly amid tall grass. Leaves tell-tale pellets in its territories. **DIET** Not well known, but mostly feeds on grass and herbs. **DISTRIBUTION** Scattered locations in Terai floodplains in Dudhwa National Park, North Bengal and Assam. Endemic, rare and endangered. **HABITAT** Tall grass of early succession stages in Terai region.

Royle's Pika ■ *Ochotona roylei*
F Ochotonidae **O** Lagomorpha

DESCRIPTION The most common pika in India. Rich russet or chestnut in colour, with some mottling of grey on head, shoulders and back. Juveniles show extensive mottling. Snout slightly arched. Shows very high geographic variation in pelage colour, hence requires further studies. Frequents grassy meadows and seeks cover under stones and boulders, darting in among them when it senses danger. **DIET** Grass and alpine herbs. **DISTRIBUTION** Himalaya from Jammu and Kashmir to Sikkim, at 2,500–4000m. Probably the only pika found at altitudes below 3,000m. **HABITAT** Subalpine and alpine meadows strewn with rocks.

Large-eared Pika ■ *Ochotona macrotis*
F Ochotonidae **O** Lagomorpha

DESCRIPTION Similar in appearance to Royle's Pika (see above), but less russet and more grey-brown in colour. Ears broad and made conspicuous by long hair emerging from them. Feet are pale. Moults before winter to adopt straw-coloured coat. Behaviour similar to that of other pikas. **DIET** Similar to that of other pikas. **DISTRIBUTION** Throughout Himalaya in alpine regions at 3,000–6,000m. **HABITAT** Alpine meadows and cold deserts strewn with rocks.

Moupin's Pika ▪ *Ochotona thibetana*
F Ochotonidae **O** Lagomorpha

DESCRIPTION Small russet-brown pika with buff underparts. Pale buff patches behind ears that are not always distinguishable from dorsal fur. In winter the colour becomes paler. Ears are moderate in size. Lives in burrows like other pikas. **DIET** Grasses, flowers and pine cones. **DISTRIBUTION** Currently known only from Sikkim at elevations of 1,800–4,100m. **HABITAT** Found at lower elevations than other sympatric pikas. Occurs in rocky areas in canopied forests and also in rocky areas above the treeline.

Ladakh Pika ▪ *Ochotona ladacensis*
F Ochotonidae **O** Lagomorpha

DESCRIPTION Large pika with sandy-brown fur and dirty-white underparts. Ears large and broad, and generally held backwards. Lives in scattered family groups and digs large holes into which it darts on sensing danger. **DIET** Grass and alpine herbs. **DISTRIBUTION** Ladakh in Jammu and Kashmir at altitudes above 4,500m. **HABITAT** Cold deserts and barren plateaus.

Indian Pangolin

■ *Manis crassicaudata*
F Manidae **O** Pholidata

DESCRIPTION Unique creature with large yellow scales covering body. Long snout and longer sticky tongue to help it catch ants and termites. Eyes small and ears like slits in skin without any ear-flap. Three long claws help it break open termite mounds. An adept climber. Rolls into a ball in self-defence. **DIET** Small insects like ants and termites. **DISTRIBUTION** Throughout mainland India except Himalaya, North-east India and desert. Now highly endangered and increasingly rare due to rampant poaching. **HABITAT** Dry deciduous forests and scrub.

Chinese Pangolin

■ *Manis pentadactyla*
F Manidae **O** Pholidata

DESCRIPTION Smaller than Indian Pangolin (see above). Smaller overlapping scales over body; face also more naked than that of Indian. Ears pronounced with prominent pinnae. Sleeps hidden in burrows during the day and emerges at night. Pups hang on to their mother's tail while out foraging. **DIET** Insects. **DISTRIBUTION** North-east India. Now critically endangered and rare due to rampant poaching. **HABITAT** Grassland, deciduous forests and bamboo groves.

Nicobar Tree Shrew
■ *Tupaia nicobarica*
F Tupaiidae **O** Scandentia

DESCRIPTION Uniform reddish-brown tree shrew with pale underparts, thereby closely resembling the allopatric Malay Tree Shrew (see p. 112). Purely arboreal and diurnal. Follows mixed hunting groups and maintains a strong foraging association with Greater Racket-tailed Drongo *Dicrurus paradiseus*. **DIET** Insects. **DISTRIBUTION** Little and Great Nicobar Islands. **HABITAT** Littoral forests and edges of rainforests.

Madras Tree Shrew ■ *Ananthana ellioti*
F Tupaiidae **O** Scandentia

DESCRIPTION Meek, squirrel-like animal; chocolate-brown with pale eye-ring and characteristic white shoulder-stripe. Snout acutely pointed. Shows slight geographic variation in pelage – Central Indian forms are more reddish, while those from dry forests of South India are more sandy-grey. Moves about actively. **DIET** Insects. **DISTRIBUTION** Peninsular India up to Jharkhand. **HABITAT** Dry deciduous forests; often in rocky areas.

Malay Tree Shrew ▪ *Tupaia belangeri*
F Tupaiidae **O** Scandentia

DESCRIPTION The *Tupaia* tree shrews are more mouse-like in appearance than the squirrel-like *Anathana*. Body uniform coffee-brown and slightly paler below. Long, furry tail is as long as body. Often found scurrying on the ground and can be mistaken for a squirrel at first sight. **DIET** Insects. **DISTRIBUTION** North-east India and Eastern Himalaya from Sikkim to Arunachal Pradesh. **HABITAT** Primary and secondary evergreen forests, bamboo groves, plantation and gardens.

Grey Musk Shrew ▪ *Suncus murinus*
F Soricidae **O** Eulipotyphla

DESCRIPTION Common and familiar animal of houses. Grey-brown above and slightly paler below; paler in arid regions of north-west. Among the largest shrews; similar in size to House Mouse (see p. 128). Common name derives from strong, musky odour it leaves behind. Highly vocal, emitting high-pitched squeaks. Often persecuted due to its mousey appearance, but not a pest like the House Mouse as it rids houses of insect pests. **DIET** Insects. **DISTRIBUTION** Throughout India. **HABITAT** Sewers and human habitation.

Indian Hedgehog (Pale Hedgehog) ▪ *Paraechinus micropus*
F Erinaceidae **O** Eulipotyphla

DESCRIPTION Masked face with greyish-white hairs on cheeks and forehead. Spine pale sandy-brown in colour, with a grizzled appearance. Smaller ears and limbs, hence more compact than the sympatric Collared Hedgehog (see p. 114). Roosts in burrows or under *Zizyphus* clumps. Becomes torpid in unfavourable conditions. When threatened it rolls itself into a ball. **DIET** Insects, worms, lizards, mice and birds' eggs; also *Zizyphus* berries and tubers of other plants. **DISTRIBUTION** Arid North-west India from Rajasthan to western Madhya Pradesh and Uttar Pradesh. **HABITAT** Arid and rocky areas and thorn forests.

Madras Hedgehog ▪ *Paraechinus nudiventris*
F Erinaceidae **O** Eulipotyphla

DESCRIPTION Previously a subspecies of Indian Hedgehog (see above). The only hedgehog in South India. Similar in appearance to Indian Hedgehog, but has ruddy spines. Masked appearance, and small ears and limbs. Roosts in dens and burrows. When threatened it rolls itself into a ball. An endemic, little-known species. **DIET** Insects and probably even berries of shrubs. **DISTRIBUTION** Parts of southern Andhra Pradesh, northern Tamil Nadu and Kerala. **HABITAT** Arid scrub forests and rocky hills.

Collared Hedgehog ■ *Hemiechinus collaris*
F Erinaceidae **O** Eulipotyphla

DESCRIPTION Distinguished from the sympatric Indian Hedgehog (see p. 113) by its dark spines, black belly and tail, and long ears. Roosts in burrows or dens, and also becomes torpid during unfavourable conditions. When threatened it rolls itself into a ball. **DIET** Insects, worms, lizards, mice and birds' eggs; also *Zizyphus* berries and tubers of other plants. **DISTRIBUTION** Arid north-west from Rajasthan to western Madhya Pradesh and Uttar Pradesh. **HABITAT** Arid and rocky areas, and thorn forests.

Indian Porcupine ■ *Hystrix indica*
F Hystricidae **O** Rodentia

DESCRIPTION Large and robust rodent, armed profusely with long, black-and-white quills and an under-armature of black bristles. Long black crest. Individuals from central

and southern India often russet in colour – a form called the 'Red Porcupine'. Roosts in dens and caves, or makes a network of interconnected burrows. Roost often strewn with bones, which it gnaws at (also doing the same with fallen antlers of deer) to strengthen its quills. Leaves behind tell-tale cigar-shaped droppings. Nocturnal. **DIET** Vegetables, grains, fruits, roots and tubers. **DISTRIBUTION** Throughout mainland North-east India. **HABITAT** Rocky hillsides, dry and moist forests, and dry scrub.

Himalayan Crestless Porcupine ■ *Hystrix brachyura*
F Hystricidae **O** Rodentia

DESCRIPTION Smaller than Indian Porcupine (see opposite), with shorter quills that are double banded with black and white. Also distinguished from Indian by very short or absent crest. Ranges generally do not overlap. Like Indian, lives in caves, dens and burrows. Due to its short quills, does not produce strong rattling sound when provoked. Nocturnal. **DIET** Vegetables, grains, fruits, roots and tubers. **DISTRIBUTION** North Bengal, Sikkim and North-east India. **HABITAT** Moist forests and edges with rocky outcrops.

Brush-tailed Porcupine ■ *Atherurus macrourus*
F Hystricidae **O** Rodentia

DESCRIPTION Unusual, rare and little-known porcupine. Smallest porcupine in India, with very short, grey-brown, spiny quills all over body similar to fur. Tip of tail brush-like with a tangle of long quills. Roosts in underground burrows during the day. Nocturnal. **DIET** Vegetables, grains, fruits, roots and tubers; also an effective seed predator. **DISTRIBUTION** North-east India; currently known only from Assam, Arunachal Pradesh and Meghalaya. **HABITAT** Dense rainforests with thick growth of bamboo and rattans.

Himalayan Marmot ■ *Marmota himalayana*
F Sciuridae **O** Rodentia

DESCRIPTION Large, stocky and robust rodent that lives in the Trans-Himalaya. Coat short but coarse, and uniform tawny-brown in colour, often with black blotches but never a

dark, saddle-like patch like that of Long-tailed Marmot (see below). Lives in large groups that excavate deep, interconnected burrows. A sentry may stand guard on its haunches and shriek in alarm when danger threatens, at which its comrades dart into burrows. Hibernates in winter. **DIET** Grasses, herbs, grains, roots and tubers; also all sorts of junk food from unruly tourists in Ladakh. **DISTRIBUTION** Throughout Himalaya at 3,500–5,200m. **HABITAT** Alpine meadows, grassland and cold deserts.

Long-tailed Marmot ■ *Marmota caudata*
F Sciuridae **O** Rodentia

DESCRIPTION Similar in size to Himalayan Marmot (see above), but easily distinguished by its long, bushy tail. Coat is rich golden (though it can become drab brown in some

individuals or in certain seasons), with long fur compared with coarse tawny-brown coat of Himalayan. Distinctive, saddle-like black patch from nape to rump. Similar in habits to Himalayan, but less common. **DIET** Grasses, herbs, grains, roots and tubers. **DISTRIBUTION** Zanskar in Ladakh and Gilgit in Jammu, and Kashmir and Sikkim, at 3,200–5,000m. **HABITAT** Alpine meadows with dwarf junipers and grasses.

Indian Giant Squirrel ■ *Ratufa indica*
F Sciuridae **O** Rodentia

DESCRIPTION Large and robust squirrel. Fur generally reddish above and buffy below. Shows high geographic variation in pelage colour. Individuals from Northern Western Ghats are purely chestnut above, while those from Central and Southern Western Ghats are generally black with maroon on the back. Arboreal. Announces its presence with loud alarm calls comprising a set of shrieking notes and barks. Builds a nest (the size of a raptor's nest) during breeding season. **DIET** Flowers, fruits and foliage of a variety of forest trees. **DISTRIBUTION** Western Ghats and scattered localities in Eastern Ghats in Andhra Pradesh, and Odisha and Satpudas in Madhya Pradesh and Maharashtra. **HABITAT** Forest dependent; dry and moist deciduous forests.

Malayan Giant Squirrel ■ *Ratufa bicolor*
F Sciuridae **O** Rodentia

DESCRIPTION Slightly smaller than Indian Giant Squirrel (see above). An elegant species, dark brown or black above and buffy-yellow below. Ears large and black, with tufts. Arboreal. Fairly vocal and similar in behaviour to Indian Giant. **DIET** Flowers, fruits and foliage of a variety of forest trees. **DISTRIBUTION** North Bengal and North-east India. **HABITAT** Low-lying and montane moist deciduous and semi-evergreen forests.

Grizzled Giant Squirrel

▪ *Ratufa macroura*
F Sciuridae **O** Rodentia

DESCRIPTION Smallest giant squirrel in India. Pelage varies from grey to dark brown above and pale below. Dorsal pelage interspersed with pale and dark hairs, giving it a grizzled appearance (especially at the tail). Characteristic black crown and tip of snout, and pale pink lips and nose. Vocal. Makes a nest during breeding season in late summer. **DIET** Flowers and fruits of dry-scrub forest plant species. **DISTRIBUTION** Isolated pockets of South India between dry rain shadow of Western Ghats and Eastern Ghats. **HABITAT** Riverine patches in dry-scrub forests of rain-shadow regions of Western Ghats.

Red Giant Flying Squirrel ▪ *Petaurista petaurista*
F Sciuridae **O** Rodentia

DESCRIPTION Large flying squirrel with red coat. Males chestnut-red and females rufous-brown. There is very high geographic variation and at least two races of the species are recognized – one with buff underparts and the other with pure white underparts. Climbs high up on one tree, then glides from tree to tree, covering distances of up to 100m in a glide. Nasal call that is produced frequently at dusk. **DIET** Fruits and flowers; also gnaws on bark. **DISTRIBUTION** Throughout Himalaya and North-east India. **HABITAT** Oak, pine and deodar forests in Western Himalaya, and tropical broadleaved forests in Eastern Himalaya and North-east India.

Indian Giant Flying Squirrel ■ *Petaurista philippensis*
F Sciuridae **O** Rodentia

DESCRIPTION The most common flying squirrel on the peninsula. Coat varies from rufous to grey grizzled with white. Pale grey underparts and rufous patagium. High geographic variation; individuals from south are darkest and often have rufous upperparts. Head concolourous to body. Most active at dusk. **DIET** Fruits and flowers; also gnaws on bark. **DISTRIBUTION** Scattered localities in undisturbed forests all over the peninsula. **HABITAT** Undisturbed, dry and moist deciduous forests with tall stands of trees.

Bhutan Giant Flying Squirrel ■ *Petaurista nobilis*
F Sciuridae **O** Rodentia

DESCRIPTION Large, elegant flying squirrel; tawny-golden in colour with two parallel chestnut stripes running from snout to vent, and yellow patch on crown. Patagium chestnut in colour; long rufous tail tipped with black, and feet are maroon. Largely crepuscular and nocturnal. **DIET** Fruits and flowers; gnaws on bark and also licks salt at salt licks. **DISTRIBUTION** Eastern Himalaya from Sikkim to Arunachal Pradesh. **HABITAT** Montane forests of oaks, pines and rhododendrons.

Grey-headed Flying Squirrel ■ *Petaurista caniceps*
F Sciuridae **O** Rodentia

DESCRIPTION A little-known, rather small flying squirrel from Eastern Himalaya. Body rich ochraceous or chestnut; head contrastingly smoky-grey. Large ears. Body striped or mottled with black. Very little known of its habits. **DIET** Similar to that of other flying squirrels. **DISTRIBUTION** Eastern Himalaya from Sikkim to Arunachal Pradesh. **HABITAT** Montane forests of oaks, pines and rhododendrons.

Parti-coloured Flying Squirrel ■ *Hylopetes alboniger*
F Sciuridae **O** Rodentia

DESCRIPTION Small, docile-looking flying squirrel with grey-brown upperparts and pure white underparts. Tail also grey to brown, with bold mid-dorsal stripe. Produces high-pitched call. **DIET** Similar to that of other flying squirrels. **DISTRIBUTION** Currently known from North Bengal and Sikkim, and hills of Arunachal Pradesh, Assam, Nagaland, Meghalaya and Manipur. **HABITAT** Primary and secondary tropical evergreen forests, often near human habitation.

Travancore Flying Squirrel ■ *Petinomys fuscocapillus*
F Sciuridae **O** Rodentia

DESCRIPTION Small, coffee-brown flying squirrel with yellowish underparts. Shy and uncommon animal with very little known about its habits. Nearly strictly a canopy dweller. Only sympatric with Indian Giant Flying Squirrel (see p. 119), from which it is differentiated by its small size and more delicate build. **DIET** Similar to that of other flying squirrels. **DISTRIBUTION** Western Ghats of Kerala and Tamil Nadu. **HABITAT** primary evergreen forests, forest edges and shade-grown coffee plantations.

Hoary-bellied Squirrel ■ *Callosciurus pygerythrus*
F Sciuridae **O** Rodentia

DESCRIPTION Drab brown squirrel, larger in size than the familiar striped squirrels of cities. Pale underparts and distinctive rufous tinge at bases of limbs. Face appears more robust than those of the species following Pallas's Squirrel (see p. 122) in this book. Long tail approximately as large as rest of body. Vocal, producing harsh chuckles. **DIET** Fruits, flowers and foliage. **DISTRIBUTION** Sikkim, North Bengal and plains of North-east India. **HABITAT** Montane forests in Himalaya, and lightly wooded areas and human habitation in North-east India.

Pallas's Squirrel ■ *Callosciurus erythraeus*
F Sciuridae O Rodentia

DESCRIPTION Robust squirrel, slightly larger than Hoary-bellied Squirrel (see p. 121). Upperparts olive-brown with a grizzled appearance. Underparts strikingly red, but often not seen easily in the field. Tail lightly ringed at tip. Vocal, producing a deep coughing sound like that of Great Pied Hornbill *Buceros bicornis*. **DIET** Fruits, flowers and foliage. **DISTRIBUTION** North-east India. **HABITAT** Primary and secondary forests. Not uncommon in dense forests. Avoids human habitation.

Himalayan Striped Squirrel ■ *Tamiops macclellandi*
F Sciuridae O Rodentia

DESCRIPTION Dainty squirrel, superficially resembling the familiar palm squirrels. Head small and rounded, with white eye-ring. Broad black stripe at the back and two thin white stripes on flanks bordered with black. Hyperactive and moves about swiftly amid dense foliage. **DIET** Small fruits, berries and flowers. **DISTRIBUTION** Sikkim, North Bengal and North-east India. **HABITAT** Montane forests in Himalaya and dense, low-lying evergreen forests in North-east India.

Orange-bellied Himalayan Squirrel ■ *Dremomys lokriah*
F Sciuridae **O** Rodentia

DESCRIPTION Drab squirrel, superficially like a small Hoary-bellied Squirrel (see p. 121). Face is small and meeker than that of Hoary-bellied Squirrel, often with some rufous at snout. Unlike the suggestion in its name, orange belly is barely visible. Belly faint rufous or orange, and never bright enough to be conspicuous. Shy species, moving about actively. **DIET** Fruits, flowers and foliage. **DISTRIBUTION** Sikkim, North Bengal and hills of North-east India. **HABITAT** Montane forests and foothills.

Three-striped Palm Squirrel ■ *Funambulus palmarum*
F Sciuridae **O** Rodentia

DESCRIPTION Familiar squirrel, commensal of humans in peninsular India. Sooty-brown to black above, with three bold, parallel white stripes on back. Two broken and faint stripes visible on flanks. Tail as long as rest of body, and bushy with soft hair. Highly vocal, with a shrill, bird-like *tee…tee…tee* call repeated endlessly. **DIET** Berries and other fruits, flowers and small insects; also known to raid birds' nests. **DISTRIBUTION** Peninsular India south of Madhya Pradesh and Jharkhand. **HABITAT** Gardens, groves, human habitation and lightly wooded areas.

Five-striped Palm Squirrel ■ *Funambulus pennantii*
F Sciuridae **O** Rodentia

DESCRIPTION Familiar, commensal species on Northern Plains. Generally paler than Three-striped Palm Squirrel (see p. 123), but as dark as it in some parts of its range. Light sandy-brown above with three bold, parallel white stripes on back and two on flanks. Where ranges overlap with those of Three-striped, best separated by its trilling, bird-like call, which is given out with frenzied jerking of the tail. **DIET** Berries and other fruits, flowers and small insects; also known to raid birds' nests. **DISTRIBUTION** North India from south of Jammu and Kashmir to Maharashtra. Ranges overlap with those of Three-striped in Maharashtra, Andhra Pradesh and Odisha. Introduced in Andaman and Nicobar Islands. **HABITAT** Gardens, groves, human habitation and lightly wooded areas.

Jungle Striped Squirrel ■ *Funambulus tristriatus*
F Sciuridae **O** Rodentia

DESCRIPTION Larger and slightly bulkier than the previous striped squirrels, with a ruddy colouration. Dorsal fur dark, like that of Three-striped Palm Squirrel (see p. 123), but face, rump and hindquarters are rufous. Similar in behaviour to other striped squirrels. **DIET** Similar to that of other striped squirrels. **DISTRIBUTION** Endemic to Western Ghats from Dangs in Gujarat to Kerala. **HABITAT** Moist deciduous hill forests.

Dusky Striped Squirrel ■ *Funambulus sublineatus*
F Sciuridae **O** Rodentia

DESCRIPTION Uncommon, diminutive squirrel, darker than sympatric Three-striped and Jungle Striped Squirrels (see p. 123). The three stripes on its back are diffused in its dark colouration. Tail as long as body. Follows mixed species flocks of birds in rainforests. **DIET** Berries and other fruits, and insects. **DISTRIBUTION** Endemic to Western Ghats in Karnataka, Kerala and Tamil Nadu. **HABITAT** Undergrowth in moist deciduous, tropical rainforests and sholas.

Stoliczka's Mountain Vole ■ *Alticola stoliczkanus*
F Muridae **O** Rodentia

DESCRIPTION Small rodent found in higher reaches of Himalaya. Mainly fossorial and lives in burrows. Body cylindrical, muzzle short and rounded, and tail half the length of body. Bright rufous-brown above with slaty-grey or white underparts. Lives close to its burrows and darts into them at the slightest threat. Might be confused with a pika, but small size and presence of a tail are distinctive. Chiefly diurnal. **DIET** High-altitude herbs, grass and grains. **DISTRIBUTION** Northern Ladakh and Sikkim. **HABITAT** Alpine meadows above 3,000m in close proximity to waterbodies where the mud is loose and conducive for burrowing.

Indian Gerbil ▪ *Tatera indica*
F Muridae **O** Rodentia

DESCRIPTION Gerbils and jirds are also called 'antelope-rats', as they have large hind limbs and move about in a series of leaps and bounds. They are distinguishable from mice and rats by a long (longer than the body), hairy tail (naked in mice and rats) that

ends in a tassel. This species is larger than Indian Desert Jird (see below). It is biscuit coloured, with a longer tail that is half cream and half black. Feet are pale and it holds a more erect posture. Lives in networks of interconnected burrows. Chiefly nocturnal, but also active during the day. **DIET** Grains, roots and grass, but also grubs and birds' eggs; partial to fruits of prickly pear in dry season. **DISTRIBUTION** Throughout mainland India, excluding High Himalaya and North-east India. **HABITAT** Desert and open semi-desert, fields and fallow land.

Indian Desert Jird
▪ *Meriones hurrinae*
F Muridae **O** Rodentia

DESCRIPTION Smaller than Indian Gerbil (see above), with shorter tail that is uniformly brown in colour and ends in tassel of black hair. Grey-brown above and paler below. Incisors are orange. Much more gregarious than Indian Gerbil, and lives in wider interconnected burrows. When alarmed, produces a drumming sound by stamping the hind limbs. Chiefly diurnal. **DIET** Seeds, tubers, grass, leaves, flowers and nuts of *Salvadora*. **DISTRIBUTION** Arid deserts of Western Gujarat and Rajasthan. **HABITAT** Barren, fallow land and semi-desert.

Malabar Spiny Dormouse ■ *Platacanthanomys lasiurus*
F Muridae **O** Rodentia

DESCRIPTION Unique rodent with no close relatives. A small mouse with spiny, light brown fur above and cream venter. Tail as long as body, with a bushy tip. Chiefly arboreal; roosts in tree hollows during the day. Probably builds a nest during the breeding season. **DIET** Seeds and berries; also known to be a pest of peppercorns. **DISTRIBUTION** Endemic to Western Ghats south of Shimoga. **HABITAT** Dense evergreen forests and shaded plantations.

Long-tailed Tree Mouse ■ *Vandeleuria oleracea*
F Muridae **O** Rodentia

DESCRIPTION Beautiful little mouse with fur that is soft and fawn above, and cream-white below. Eyes large, giving it a docile appearance. Tail longer than body. Exclusively arboreal and moves about swiftly, often using its long tail to grip branches. Females known to build a large nest out of leaves, bamboo blades or ribbon grass for their young. **DIET** Fruits, buds and tender shoots. **DISTRIBUTION** Throughout mainland India. **HABITAT** Dry and moist deciduous forests, bamboo forests and dry-scrub forests on hill slopes.

Long-tailed Field Mouse ■ *Apodemus sylvaticus*
F Muridae **O** Rodentia

DESCRIPTION Small mouse the size of a House Mouse (see below), found in Himalaya. Yellowish-brown in colour. Superficially resembles House Mouse, but on a closer look white feet are diagnostic and tail is bicoloured. Ears large and rounded. Upper incisors faint orange in colour. Roosts in burrows during the day and caches food for winter. Chiefly nocturnal. **DIET** Fruits, seeds and grass. **DISTRIBUTION** Throughout Himalaya at 1,800–3,500m. **HABITAT** Montane forests and alpine meadows.

House Mouse ■ *Mus musculus*
F Muridae **O** Rodentia

DESCRIPTION Probably the most familiar species in this book, the House Mouse is the common mouse of homes. In build and appearance it is a small version of the common Black Rat, or House Rat (see p. 131). Fur varies from grey to brown, and shows high geographical variation. Tail longer than body. Hyperactive, running swiftly on the ground and equally at ease while scaling walls of houses. A prolific breeder. **DIET** Omnivorous and eats practically everything. **DISTRIBUTION** Throughout India; also probably introduced in Andaman and Nicobar Islands. **HABITAT** Commensal of humans, rarely moving far from human habitation.

Indian Bush Rat ■ *Gollunda ellioti*
F Muridae **O** Rodentia

DESCRIPTION Similar in appearance to Black Rat (see p. 131), this species is slightly smaller. Fur yellowish-brown speckled with black. Head rather vole-like, rounded with a blunt muzzle. Ears very large and conch-like. Feet mostly naked. Lives in burrows during the day. Nocturnal. **DIET** Grass, grains, seeds and crops; partial to seeds of *Lantana*. **DISTRIBUTION** Peninsular India up to Haryana in north, and Western Assam in east. **HABITAT** Crop fields, scrub country and dry forests.

Large Bandicoot Rat ■ *Bandicota indica*
F Muridae **O** Rodentia

DESCRIPTION Largest rat in India, measuring about 300mm in body length. Large and bulky with dark grey to black fur. Tail naked and equally long. A commensal of humans, though not as common as Black Rat (see p. 131). Does not move very fast due to its large size. Aggressive, and erects its hair and grunts when excited. Nocturnal. **DIET** Omnivorous; eats anything available. **DISTRIBUTION** Throughout India. **HABITAT** Human habitation and crop fields.

Lesser Bandicoot Rat ▪ *Bandicota bengalensis*
F Muridae **O** Rodentia

DESCRIPTION Robust rodent measuring about 150mm in body length. Rounded face, round, pinkish ears and short, broad muzzle. Coat greyish-brown speckled with buff. Erects its long hair and grunts when excited. Lives in interconnected burrows where it leaves behind a pile of fresh earth resembling a molehill. **DIET** Omnivorous; eats anything available. **DISTRIBUTION** Throughout mainland India (up to 3,500m in Himalaya). **HABITAT** Human habitation, crop fields and vegetation near wetlands.

Short-tailed Bandicoot Rat ▪ *Nesokia indica*
F Muridae **O** Rodentia

DESCRIPTION Large rodent similar in size to Lesser Bandicoot Rat (see above). Dull brown in colour with paler underparts. Tail rather short and dark. Very similar to Lesser Bandicoot Rat, but not as much of a commensal. Lives in interconnected burrows. **DIET** Generalist and omnivorous, but mostly eats grass and crops, thereby causing extensive damage in crop fields. **DISTRIBUTION** Northern India. **HABITAT** Widespread, including in crop fields and damp places.

Black Rat (House Rat) ■ *Rattus rattus*
F Muridae **O** Rodentia

DESCRIPTION The most common rat in the world. Coat highly variable and many subspecies are recognized. Dorsal pelage generally dark brown to black. Some forms in hills have long and soft fur. A commensal of man, highly adaptable and a successful colonizer. **DIET** Anything from human refuse, vegetables and meat, to carrion. **DISTRIBUTION** Throughout India. **HABITAT** Human habitation and forests.

Brown Rat ■ *Rattus norvegicus*
F Muridae **O** Rodentia

DESCRIPTION Common and cosmopolitan rat throughout the world. Very similar to Black Rat (see above), but with tail that is marginally shorter than body. Light or dark brown on back and paler below. Largely spread throughout the world on ships, hence chiefly distributed in coastal cities and islands in India. **DIET** Anything. **DISTRIBUTION** Coastal cities, where it may outnumber Black Rat. **HABITAT** Human habitation; frequents drains and sewers.

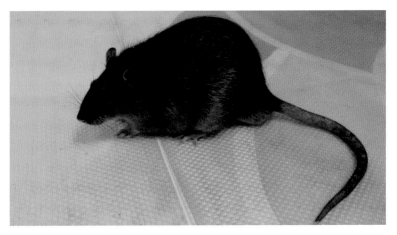

Indian Flying Fox

▪ *Pteropus giganteus*
F Pteropodidae **O** Chiroptera

DESCRIPTION Very large fruit bat with average forearm length of 168mm, chestnut-brown head and large black ears. Belly paler than rest of body. Wings huge and black with long thumbs. Gregarious and roosts during the day in open or shaded trees in villages and towns. Fans itself constantly with its wings on hot summer days. **DIET** Fruits and nectar of native and exotic trees. Some authorities believe that it only bites fruits to suck out the sap, discarding the remainder. **DISTRIBUTION** Widespread; found throughout India except parts of arid north-west and at high elevations of Himalaya. **HABITAT** Roosts near human habitation, mostly on old trees; rarely ventures into dense forest.

Fulvous Fruit Bat

▪ *Rousettus leschenaultii*
F Pteropodidae **O** Chiroptera

DESCRIPTION Medium-sized bat with average forearm length of 82mm; largely fulvous brown or grey in colour. Juveniles more grey than adults. Snout fairly long. Highly gregarious and roosts in very large colonies in caves, disused buildings and heritage monuments; rarely on trees. Roost smells strongly like dry ammonia. **DIET** Fruits and nectar of a variety of species; consumes leaves when fruits are unavailable. **DISTRIBUTION** Throughout India except desert and High Himalaya, and Andaman and Nicobar Islands. **HABITAT** Variety of habitats including human habitation, dry deciduous forests and dense evergreen forests.

Greater Short-nosed Fruit Bat ■ *Cynopterus sphinx*
F Pteropodidae **O** Chiroptera

DESCRIPTION Medium-sized bat with average forearm length of 70mm; smaller than Fulvous Fruit Bat (see opposite), with short and broad snout. Pelage varies from grey to rich rufous and orange; adults generally darker than juveniles. Prominent pale fingers on black wings, and pale borders to ears in adults. Roosts under fronds of palm leaves in small colonies or harems; also infrequently in disused buildings and caves. Males make a tent by biting palm leaves during breeding season. **DIET** Fruits and nectar of several neighbourhood plants like guava, mango, date and fishtail palms, chiku and similar. **DISTRIBUTION** Widespread and very common. Found throughout India, including on Andaman and Nicobar Islands; absent from arid north-west and High Himalaya. **HABITAT** Mostly commensal to man, and common around human habitation in gardens, groves and orchards; also occurs in lightly wooded areas, but uncommon in moist forests.

Lesser Short-nosed Fruit Bat ■ *Cynopterus brachyotis*
F Pteropodidae **O** Chiroptera

DESCRIPTION Smaller version of Greater Short-nosed Fruit Bat (see above), with smaller average forearm length of 60mm and smaller ears. Lacks pale fingers and ear border of Greater. Not much is known about its roosting behaviour in India, but it is known to roost similarly in South-east Asia. **DIET** Fruits and nectar of several plant species. **DISTRIBUTION** Reported from Western Ghats south of Maharashtra and Eastern Ghats of Andhra Pradesh, North-east India, Andaman and Nicobar Islands, and Sri Lanka. **HABITAT** Unlike Greater, uncommon near human habitation on Indian subcontinent, and restricted to moist forests.

Cave Nectar Bat ■ *Eonycteris spelaea*
F Pteropodidae **O** Chiroptera

DESCRIPTION Medium-sized bat with average forearm length of 72mm. Superficially similar to Fulvous Fruit Bat (see p. 132) due to long snout and similar pelage, but marginally smaller and lacks claw on first finger. Pelage varies from grey to dark brown.

Tongue long and often held out of mouth. Like Fulvous Fruit Bat, roosts in large colonies in caves, and the two species often share a roost. Roost smells strongly like dry ammonia. **DIET** Predominantly nectar and pollen from night-flowering plants; also fruits. In South-east Asia it is an important pollinator of durian. **DISTRIBUTION** Reported from parts of Uttarakhand, Western Ghats, Eastern Ghats, North-east India, and Andaman and Nicobar Islands. Found up to 1,000m in Himalaya. **HABITAT** Mainly found in forested areas, and less cosmopolitan than Fulvous Fruit Bat.

Salim Ali's Fruit Bat ■ *Latidens salimalii*
F Pteropodidae **O** Chiroptera

DESCRIPTION Medium-sized bat with average forearm length of 67mm. Resembles Short-nosed Fruit Bat (see p. 133) but its pelage is dark brown to black, and it lacks pale fingers, ear margins and short tail of Short-nosed. Roosts in natural caves in moderately large colonies. **DIET** Fruits, but not much is known about its diet. **DISTRIBUTION** Endangered endemic bat. Restricted in range; known only from High Wavy Mountains, Anamalais and Kalakkad-Mundanthurai Tiger Reserve in Tamil Nadu and Periyar TR in Kerala. **HABITAT** Dense montane evergreen forests and coffee estates.

Greater Mouse-tailed Bat ▪ *Rhinopoma microphyllum*
F Rhinopomatidae O Chiroptera

DESCRIPTION Mouse-tailed bats are characterized by their long, whip-like tails. This species has an average forearm length of 70mm, with forearm being longer than tail. Face and ears are naked; nose has a groove. Pelage varies from grey to rufous. Roosts gregariously in caves, abandoned forts, tunnels and wells. Hangs on all fours on vertical walls; when disturbed crawls in a crab-like fashion. Roost smells strongly like dry ammonia. **DIET** Various insect orders like beetles, moths, termites (seasonally) and cockroaches. **DISTRIBUTION** Dry, arid and semi-arid parts of north and north-west India; predominantly in Rajasthan, Delhi, Haryana, parts of Gujarat and Madhya Pradesh. **HABITAT** Dry and arid regions; feeds high above the ground over dry desert and scrub country.

Lesser Mouse-tailed Bat ▪ *Rhinopoma hardwickii*
F Rhinopomatidae O Chirotera

DESCRIPTION Smaller than Greater Mouse-tailed Bat (see above), with average forearm length of 58mm; forearm shorter than tail. Otherwise similar in appearance and behaviour to Greater. **DIET** Various insect orders like beetles, moths, termites and cockroaches. **DISTRIBUTION** Throughout India, but absent from Himalaya, Western Ghats, North-east India, and Andaman and Nicobar Islands. **HABITAT** Dry and arid regions, scrub country and dry deciduous forests.

Egyptian Free-tailed Bat ▪ *Tadarida aegyptiaca*
F Molossidae O Chiroptera

DESCRIPTION Medium-sized bat with average forearm length of 50mm. Free-tailed bats have peculiar naked faces with wrinkled lips; ears project forwards and ahead of eyes and are not connected to each other; snout rather broad and pig-like when seen from the front. Pelage mostly uniform dark buffy to clove-brown, slightly paler ventrally. Short, thick tail emerges free of body. Roosts in crevices and slits in caves, and abandoned monuments and heritage buildings. Swift, gliding flight. Smell unpleasant to humans. **DIET** Various types of insect like beetles, moths, crickets and flies. Feeds high above the ground. **DISTRIBUTION** Peninsular India; currently reported from Gujarat, Western Maharashtra, Karnataka, Kerala, Tamil Nadu and Andhra Pradesh, and one record from Lakshadweep, but may be more widespread. **HABITAT** Dry and semi-arid areas.

Wrinkle-lipped Bat ▪ *Chaerephon plicata*
F Molossidae O Chiroptera

DESCRIPTION Smaller than Egyptian Free-tailed Bat (see above), with average forearm length of 46mm. Similar in appearance to Egyptian Free-tailed, but ears are connected to

each other by membrane over forehead. Pelage similar to that of Egyptian Free-tailed, but slightly darker above, and there is a contrast between dark dorsum and white or creamish venter. Gregarious, roosting in large colonies in high-walled limestone caves, on steep rock faces and occasionally in buildings. Swift and gliding flight. Smell unpleasant to humans. **DIET** Variety of insects, probably similar to that of Egyptian Free-tailed. **DISTRIBUTION** Scattered localities all over India except desert, Himalaya, and Andaman and Nicobar Islands; possibly under-reported. **HABITAT** Feeding habitat probably wetter and denser forests than those used by Egyptian Free-tailed; also feeds over large rivers in forests and gorges.

Wroughton's Free-tailed Bat ■ *Otomops wroughtonii*
F Molossidae **O** Chiroptera

DESCRIPTION Largest of free-tailed bat family in India, with average forearm length of 64mm; more robust than former free-tailed bats, and ears connected over forehead. Pelage sooty or black above, and cream or white below, with a distinctive white collar. Roosts in clumps of 50–100 individuals in cracks and crevices in limestone caves. Swift flight. **DIET** Unknown, but probably similar to that of other free-tailed bats. **DISTRIBUTION** Until 2014 known from only one location near Belgaum in North Karnataka; recently discovered in Meghalaya. **HABITAT** Unknown; roosts situated in dense evergreen forests; possibly similar to that of Wrinkle-lipped Bat (see opposite).

Long-winged Tomb Bat ■ *Taphozous longimanus*
F Emballonuridae **O** Chiroptera

DESCRIPTION Medium-sized bat with average forearm length of 60mm. Tomb bats are characterized by their mouse-like, triangular faces, small ears and short tails, which emerge partly free of the interfemoral membrane. Species are rarely distinguishable without handling. Pelage of this species is of varying shades of brown, from tawny to cinnamon. Fingers larger than those of other tomb bats, and wings attached to ankles of hindlegs. Roosts in groups of up to 50 individuals in caves, crevices, eaves of houses and large, hollow tree trunks. Flies very fast. Sexes roost separately in non-breeding season. **DIET** Insects like beetles, termites, cockroaches and moths. **DISTRIBUTION** Widespread in peninsular India, with one record from Darjeeling; not reported from North India. **HABITAT** Feeds around human habitation and open forests; generally flies high above the ground.

Naked-rumped Tomb Bat ■ *Taphozous nudiventris*
F Emballonuridae **O** Chiroptera

DESCRIPTION Larger than Long-winged Tomb Bat (see p. 137), with average forearm length of 74mm. Adults differentiated from other tomb bats by their naked rumps. Pelage mostly dark brown above with darker patches; greyish-brown ventrally. Roosts in large-walled caves and monuments; also in cracks and crevices in disused buildings. Sexes roost separately in non-breeding season. **DIET** Similar to that of other tomb bats. **DISTRIBUTION** Gujarat, Delhi, Uttar Pradesh and Bihar. **HABITAT** Arid and semi-arid areas; probably feeds over scrub country and open forests.

Black-bearded Tomb Bat ■ *Taphozous melanopogon*
F Emballonuridae **O** Chiroptera

DESCRIPTION Medium-sized bat with average forearm length of 64mm, but with shorter fingers than Long-winged Tomb Bat (see p. 137). Adult males distinguished by prominent black beard on chin. Pelage varies from tawny to dark brown. The most familiar tomb bat; roosts gregariously in caves, tunnels, forts, monuments and coves on the seashore. Roosts smell strongly like dry ammonia. Sexes roost separately in non-breeding season;

generally carries pups during monsoons. Mostly hangs on vertical walls and scurries in a crab-like manner when disturbed. **DIET** Similar to that of other tomb bats. **DISTRIBUTION** Widespread in dry and arid regions of the peninsula, and not common in North India; also occurs on Andaman and Nicobar Islands. **HABITAT** Dry, arid and semi-arid areas; feeds high above the ground.

Pouch-bearing Bat

■ *Saccolaimus saccolaimus*
F Emballonuridae **O** Chiroptera

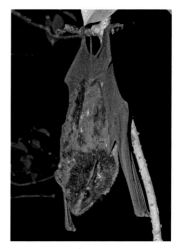

DESCRIPTION Medium-sized bat with average forearm length of 64mm. Darkest of all tomb bats found in India; dark brown to black above. Favoured roosts in South-east Asia known to be tall, hollow palm trees; rarely roosts in buildings. Roosts individually or in pairs. A little-known species. **DIET** Not well known, but probably similar to that of other tomb bats. **DISTRIBUTION** Patchily distributed in moist areas like Western Ghats, Eastern Ghats, North-east India and Andaman Islands. Probably common in East and North-east India. **HABITAT** The only tomb bat in India that prefers moist areas; recorded from rubber plantations, and moist deciduous and evergreen forests.

Greater False Vampire ■ *Megaderma lyra*

F Megadermatidae **O** Chiroptera

DESCRIPTION Robust bat with average forearm length of 66mm. Face looks like that of a rabbit with beady eyes and large, oval ears. Ears joined over forehead. Complex nose composed of a noseleaf that is tall with straight sides. Pelage uniform mouse-grey, slightly paler ventrally. Roosts in colonies of up to 50–100 individuals in abandoned houses, caves, temples and forts. **DIET** Carnivorous, eating insects, lizards, frogs, rodents, small birds and – more rarely – even other bats. **DISTRIBUTION** Throughout India except Himalaya, deserts, and Andaman and Nicobar Islands. **HABITAT** Dry deciduous forests; not reported from moist evergreen forests. Feeds low over the ground along trails and leaf litter.

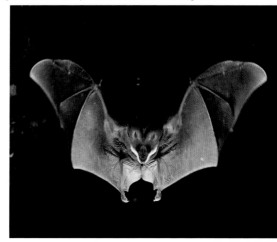

Lesser False Vampire ▪ *Megaderma spasma*
F Megadermatidae **O** Chiroptera

DESCRIPTION Robust bat, smaller than Greater False Vampire (see p. 139) and with average forearm length of 57mm. Similar in appearance to Greater, but its ears are mostly free and joined only a little at the base, and noseleaf has a heart-shaped base. Roosting behaviour similar to Greater's. Also roosts in tree hollows in mature forests. **DIET** Similar to that of Greater. **DISTRIBUTION** Moist areas like Western Ghats, parts of Eastern Ghats, North-east India and Andaman Islands. **HABITAT** Moist deciduous and evergreen forests. Feeds low over the ground along trails and leaf litter.

Greater Horseshoe Bat ▪ *Rhinolophus ferrumequinum*
F Rhinolophidae **O** Chiroptera

DESCRIPTION Medium-sized bat with average forearm length of 58mm. Fairly robust appearance. The nose, as in all horseshoe bats, is complex with a pointed sella, protruding sella and basal horseshoe. Sella broad without curves. Ears large and pointed at tips. Pelage soft and varies from grey to brown. Roosts in fairly large colonies in caves, mines, tunnels and similar places. Hibernates in winter. **DIET** Small insects like mosquitoes, lacewings, moths and spiders. **DISTRIBUTION** Throughout Himalaya at 900–3,000m. **HABITAT** Mixed forests of oaks, conifers and rhododendrons; also active over streams.

Rufous Horseshoe Bat ■ *Rhinolophus rouxii*
F Rhinolophidae **O** Chiroptera

DESCRIPTION Medium-sized bat with average forearm length of 52mm. Sella forked and rounded at tips. Soft, woolly pelage is highly variable, from grey to rufous; adults distinctly bright orange at certain times of the year. Roosts in large colonies in caves and tunnels. Becomes torpid during cold conditions. Catches insects by launching sallies from a perch like a flycatcher. **DIET** Insects like beetles, termites, mosquitoes, moths and grasshoppers. **DISTRIBUTION** Widely distributed all over India except High Himalaya, deserts, and Andaman and Nicobar Islands. Probably a complex of several species. **HABITAT** Forests; apparently most common in moist forests.

Blyth's Horseshoe Bat ■ *Rhinophus lepidus*
F Rhinolophidae **O** Chiroptera

DESCRIPTION Small bat with average forearm length of 40mm. The sella is distinctly forked and pointed at tip like a pair of scissors. Pelage generally goes from grey to brown. Roosts in large colonies in caves and tunnels, and becomes torpid during cold conditions. Catches insects on the wing and probably does not fly high. **DIET** Small insects. **DISTRIBUTION** Throughout India except High Himalaya, and Andaman and Nicobar Islands. **HABITAT** Mostly open forests and scrub.

Big-eared Horseshoe Bat ■ *Rhinolophus macrotis*
F Rhinolophidae **O** Chiroptera

DESCRIPTION Small bat with average forearm length of 42mm. The sella is slightly wavy with a broad base; horseshoe is broad and stout. Ears very large in proportion to body. Pelage mostly grey or brown. Probably roosts in small colonies in caves and tunnels. **DIET** Unknown; probably eats small insects. **DISTRIBUTION** Uncommon species. Patchily recorded at 1,500–2,500m in Himalaya, and one record from hills of Meghalaya. **HABITAT** Dense forests of oaks, conifers and rhododendrons.

Andaman Horseshoe Bat ■ *Rhinolophus cognatus*
F Rhinolophidae **O** Chiroptera

DESCRIPTION Small bat with average forearm length of 40mm; comparable in size to the allopatric Blyth's Horseshoe Bat (see p. 141), with a bifid sella with pointed tips and rounded at the centre like surgical scissors. Pelage mostly grey to grey-brown. Roosts in

small colonies in limestone caves and hollow trees. Undergoes brief torpidity in December–January. **DIET** Unknown; probably small insects, which it often hunts by 'flycatching'. **DISTRIBUTION** Endangered, uncommon and endemic species. Andaman Islands; reported from South and North Andaman Islands, and Narcondam Island. **HABITAT** Feeds in dense, wet evergreen forests along trails and clearings.

Lesser Woolly Horseshoe Bat ■ *Rhinolophus beddomei*
F Rhinolophidae **O** Chiroptera

DESCRIPTION Larger than other horseshoe bats, with forearm length of 60mm. As its name suggests, pelage is dense, woolly and sooty-black in colour. Complex nose is broad and has lappets at base. Roosts individually or in pairs in caves, abandoned houses and tree holes. Undergoes torpidity during cold weather. **DIET** Large and probably hard-bodied insects; hunts by 'flycatching'. **DISTRIBUTION** Western Ghats, Eastern Ghats and one record from North-east India. Probably under-reported. **HABITAT** Moist forests; hunts mainly in clearings in dense forests and appears territorial.

Pearson's Horseshoe Bat ■ *Rhinolophus pearsonii*
F Vespertilionidae **O** Chiroptera

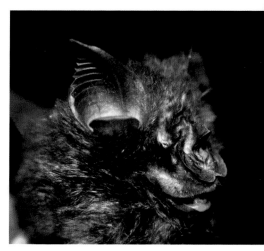

DESCRIPTION Medium-sized horseshoe bat with an average forearm of 52mm. The sella is rather 'bell-shaped', not elongated but short and deflecting downwards; as against the bluntly bifid sella of the similar-sized **Intermediate Horseshoe Bat** *Rhinolophus affinis* with which it is sympatric. Pelage goes from grey to brown. Roosts singly, in pairs or in small colonies; doesn't appear to form large colonies. Hunts like a flycatcher by launching sallies from a perch. **DIET** Unknown; probably large fluttering moths and beetles. **DISTRIBUTION** Himalayas and Northeast India **HABITAT** Undergrowth, clearings and edges of moist forests.

Woolly Horseshoe Bat

▪ *Rhinolophus luctus*
F Rhinolophidae **O** Chiroptera

DESCRIPTION Largest horseshoe bat in India, with forearm length of 75mm. Similar in appearance and behaviour to Lesser Woolly Horseshoe Bat (see p. 143), but significantly larger, and also probably allopatric (except perhaps in North-east India). **DIET** Similar to that of Lesser. **DISTRIBUTION** Throughout Himalaya (from foothills to 2,000m) and North-east India. One record from Satpura Hills in Madhya Pradesh now warrants investigation. **HABITAT** Similar to that of Lesser.

Dusky Leaf-nosed Bat ▪ *Hipposideros ater*
F Hipposideridae **O** Chiroptera

DESCRIPTION Small bat with average forearm length of 36mm. Complex noseleaf has no supplementary leaflets. Ears moderately large in proportion to face, broad at bases and slightly pointed at tips. Pelage varies from grey to rufous. Roosts in rather small colonies in caves, tunnels and abandoned houses. **DIET** Insects like small beetles, gnats and mosquitoes. **DISTRIBUTION** Peninsular India south of Maharashtra. Most common in South India. **HABITAT** Roosts in a variety of habitats, including villages; feeding habitat not well known.

Fulvous Leaf-nosed Bat ■ *Hipposideros fulvus*
F Hipposideridae **O** Chiroptera

DESCRIPTION Small bat, larger than Dusky
Leaf-nosed Bat (see opposite), with average
forearm length of 40mm. Also distinguished
by large ears, which are broadly rounded off
at tips. Nose has no supplementary leaflets.
Pelage goes from grey to bright golden-
orange. Roosts in large colonies in caves,
tunnels, abandoned houses and monuments,
often with other small horseshoe and leaf-
nosed bats. **DIET** Insects like cockroaches
and beetles; probably flies close to the
ground. **DISTRIBUTION** Throughout
India except High Himalaya, North-east
India, and Andaman and Nicobar Islands.
Common. **HABITAT** Reported from a range
of biomes, but probably most common in dry
forests and scrub.

Anderson's Leaf-nosed Bat ■ *Hipposideros pomona*
F Hipposideridae **O** Chiroptera

DESCRIPTION Small bat with average forearm length of 40mm. Very similar in
appearance to Fulvous Leaf-nosed Bat (see above), with large and blunt ears, and no
supplementary leaflets.
Not easily differentiable.
Pelage varies from grey
to golden-orange. Roosts
in caves and tunnels,
often with other small
horseshoe and leaf-nosed
bats. **DIET** Probably
similar to that of Fulvous.
DISTRIBUTION Moist
regions of Southern
Western Ghats, Eastern
Ghats, North-east
India, and Andaman
and Nicobar Islands.
HABITAT Appears to be a
species of moist forests as
compared with Fulvous.

Schneider's Leaf-nosed Bat
■ *Hipposideros speoris*
F Hipposideridae **O** Chiroptera

DESCRIPTION Medium-sized bat with average forearm length of 50mm. Differentiated from the previous leaf-nosed bats in having three pairs of supplementary leaflets (wrinkle-like structures flanking nose). Ears shorter and pointed at tips. Pelage varies from grey to brown and rufous. Roosts in caves, tunnels, abandoned forts and monuments, often with other small horseshoe and leaf-nosed bats. **DIET** Insects like beetles. **DISTRIBUTION** Peninsular India south of Maharashtra, and Gujarat. One record from Uttar Pradesh. Common and probably more widespread than current evidence suggests. **HABITAT** Variety of habitats from dry scrub to dry and moist forests. Flies along forest trails.

Horsfield's Leaf-nosed Bat ■ *Hipposideros larvatus*
F Hipposideridae **O** Chiroptera

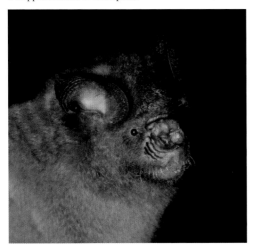

DESCRIPTION Among the largest leaf-nosed bats, with average forearm length of 62mm. Wings long, and noseleaf broad and rounded with three pairs of supplementary leaflets. Ears proportionately short and pointed at tips. Pelage mostly brown or rufous. Reported to roost in caves. **DIET** Unknown, but possibly feeds on larger insects. **DISTRIBUTION** North-east India (uncommon) **HABITAT** Not well known; forests and villages in Little Andaman. Flies fairly high while feeding.

Cantor's Leaf-nosed Bat

■ *Hipposideros galeritus*
F Hipposideridae **O** Chiroptera

DESCRIPTION Small bat with average forearm length of 47mm. Resembles Schneider's Leaf-nosed Bat (see opposite), but differs in marginally smaller size and in having two pairs of supplementary leaflets. Ears rather short and pointed at tips. Pelage mostly brown, often with dark patches on back. Roosts in small colonies in caves, tunnels and abandoned houses. **DIET** Small insects like beetles. **DISTRIBUTION** Scattered localities in peninsular India up to Madhya Pradesh. Comparatively rare; probably more widespread than currently thought. **HABITAT** Variety of habitats from dry forests and scrub to dense forests. Mostly flies along trails.

Great Himalayan Leaf-nosed Bat ■ *Hipposideros armiger*

F Hipposideridae **O** Chiroptera

DESCRIPTION Largest insectivorous bat in India, with average forearm length of 90mm. Very robust with a face that resembles that of a bulldog. Ears large, and noseleaf has four pairs of supplementary leaflets (one reduced). Pelage varies from brown to golden or rufous. Roosts in moderate to large colonies in caves. In flight can be confused with a small fruit bat. **DIET** Large insects. **DISTRIBUTION** Throughout Himalaya (up to 2,500m) and North-east India. Not uncommon. **HABITAT** Wooded areas and forests; feeds partly in the canopy and at low heights in clearings.

Kelaart's Leaf-nosed Bat ■ *Hipposideros lankadiva*
F Hipposideridae **O** Chiroptera

DESCRIPTION Second largest insectivorous bat in India, with average forearm length of 85mm. Similar in size to allopatric Great Himalayan Leaf-nosed Bat (see p. 147), but less robust. Four pairs of supplementary leaflets, but noseleaf has a considerably different shape. Pelage mostly grey-brown. Roosts in moderate to large colonies in caves and abandoned monuments. Flies high to hawk insects. **DIET** Large insects. **DISTRIBUTION** Scattered localities south of Delhi on Indian mainland. Some records from East India. **HABITAT** Wooded areas and forests; feeds around tall trees, forest clearings and streams.

Hodgson's Bat ■ *Myotis formosus*
F Vespertilionidae **O** Chiroptera

DESCRIPTION Small bat with average forearm length of 47mm. Brightly coloured, with cream or ginger-brown pelage. Wings bright orange with black patches between fingers. Genus *Myotis* is identified by its long, lance-shaped tragus (cartilage at base of ear). Despite its conspicuous colours, remains well concealed during the day among dry leaves or dense foliage. Hibernates in caves. **DIET** Small insects; not well known. **DISTRIBUTION** Throughout Himalaya (up to 3,000m), and parts of North-east India. Single records from Central India and Western Ghats of Karnataka. Uncommon and probably under-reported. **HABITAT** Unknown.

Nepalese Whiskered Bat ■ *Myotis muricola*
F Vespertilionidae **O** Chiroptera

DESCRIPTION Small bat with average forearm length of 34mm. Pelage soft and dark sooty-black in colour. Tragus long and lance shaped, as is characteristic of the genus. Roosts in small numbers in caves, foliage and tree hollows. Probably hibernates in cold regions. **DIET** Small insects; probably also aquatic insects. **DISTRIBUTION** Throughout Himalaya (up to 3,200m), and East and North-east India. Uncommon. **HABITAT** Not well studied; forests of varying types and probably streams.

Horsfield's Bat ■ *Myotis horsfieldii*
F Vespertilionidae **O** Chiroptera

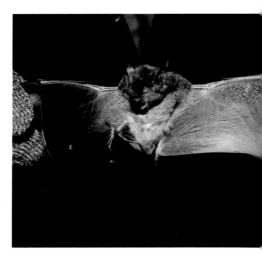

DESCRIPTION Small bat with average forearm length of 38mm. Pelage soft and dark brown to sooty-black. Ears long with long tragus. Region around small eyes generally bare. Roosts in small to large colonies in tunnels, caves, bridges and culverts. Water dependent and flies low over streams and ponds. **DIET** Aquatic insects. **DISTRIBUTION** Scattered localities throughout India except High Himalaya, and deserts. Occurs on Andaman and Nicobar Islands. Unreported from North-east India, but may well occur there. **HABITAT** Water-dependent species, favouring streams and shallow waterbodies in forests and wooded areas.

Lesser Noctule ■ *Nyctalus leisleri*
F Vespertilionidae **O** Chiroptera

DESCRIPTION Small bat with average forearm length of 44mm. Genus *Nyctalus* is identified by characteristic mushroom-shaped tragus. Pelage long and silky, and mid- to dark brown above and buffy ventrally. Ears short and rhomboidal in shape. Wings long and pointed. Roosts in tree hollows and foliage. **DIET** Variety of insects. **DISTRIBUTION** Western Himalaya from Jammu and Kashmir to Uttarakhand. **HABITAT** Generalist; occurs in forests and around open rivers from foothills to about 3,000m.

Kelaart's Pipistrelle ■ *Pipistrellus ceylonicus*
F Vespertilionidae **O** Chiroptera

DESCRIPTION Small bat with average forearm length of 35mm. Genus Pipistrellus is identified by short, blunt tragus, curved towards tip. Pelage soft, and variable in colour from brown to chestnut above, and pale to buffy-yellow below. Ears moderately large. Roosts anywhere from abandoned houses, cracks and crevices, to tree hollows and similar places. Among the earliest fliers at dusk. **DIET** Small insects like moths, flies, ants and mosquitoes. **DISTRIBUTION** Mostly peninsular India and Northeast India. **HABITAT** Generalist; also a commensal of humans. Occurs in all habitats except dense forests.

Indian Pygmy Bat ▪ *Pipistrellus tenuis*
F Vespertilionidae **O** Chiroptera

DESCRIPTION Among the smallest bats in India, with average forearm length of 27mm. Very similar to other pipistrelles, and identification is impossible without handling. Pelage, soft and variable in colour. Roosts anywhere from abandoned houses, cracks and crevices, to tree hollows and similar places. Flies out early at dusk. **DIET** Small insects. **DISTRIBUTION** Throughout India except Andaman and Nicobar Islands. **HABITAT** Generalist; found in all possible habitats except dense forests.

Kashmir Long-eared Bat ▪ *Plecotus wardi*
F Vespertilionidae **O** Chiroptera

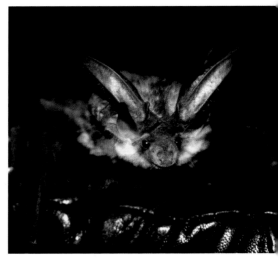

DESCRIPTION Small bat with average forearm length of 40mm. Ears are distinctly large, almost covering the whole body when in rest. Pelage soft and woolly, typically uniformly cold grey in colour. Roosts in tree hollows as well as abandoned houses or in crevices of sufficient size. Seems to fly out a little after dusk. It hunts by gleaning its prey using faint, low frequency echolocation calls. **DIET** Mostly moths and spiders. **DISTRIBUTION** High elevations (above 2,800m) in western Himalayas. **HABITAT** Occurs in alpine rhododendron forests, alpine meadows and cold desert.

Greater Yellow House Bat ▪ *Scotophilus heathii*
F Vespertilionidae **O** Chiroptera

DESCRIPTION Medium-sized bat with average forearm length of 60mm. Adults are easily identified by their size and characteristic bright yellow underparts. Tragus is sickle shaped and curved forwards. Muzzle is dark, broad and blunt. Its tail is large and enclosed in the interfemoral membrane. Roosts in abandoned caves and houses but is partial to the fronds of palm trees. Pups are born in early monsoon when insect abundance is high. **DIET** Insects like flies, beetles, and similar. **DISTRIBUTION** Throughout mainland India except High Himalaya **HABITAT** Feeds in open habitats like fields, farmland, riverine stretches and forest clearings.

Tickell's Bat ▪ *Hesperoptenus tickelli*
F Vespertilionidae **O** Chiroptera

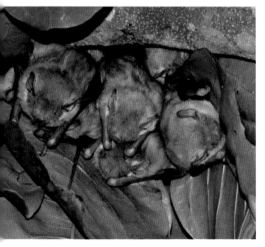

DESCRIPTION Medium-sized bat; among the largest bats in its family, with average forearm length of 55mm. Brightly coloured, with pelage that is golden-brown uniformly; forearm flesh coloured and ears dull orange. Roosts singly or in small parties in foliage, and exceedingly difficult to find at roost. Similar in appearance and roosting behaviour to Hodgson's Bat (see p. 148). **DIET** Beetles, termites and other flying insects. **DISTRIBUTION** Uncommon. Scattered locations in peninsular, East and North-east India. **HABITAT** Feeds in rather open areas like paddy fields, glades and tea plantations.

Hardwicke's Woolly Bat ■ *Kerivoula hardwickii*

F Vespertilionidae **O** Chiroptera

DESCRIPTION Small bat with average forearm length of 33mm. Small face with funnel-shaped ears, and tiny eyes set within dense pelage. Pelage dark, soft and dense. Wings are brownish. Roosts singly or in pairs in foliage in dense and moist forests; rarely in tiles of sloping roofs. In Borneo, roosts in flowers of pitcher plants. Flies out late in the day. **DIET** Small insects. **DISTRIBUTION** Scattered locations in Western and Eastern Himalaya, North-east India and Western Ghats. Rare and little-known species. **HABITAT** Dense and wet forests. Feeds in highly cluttered areas with dense undergrowth.

Eastern Bent-winged Bat ■ *Miniopterus fuliginosus*

F Miniopteridae **O** Chiroptera

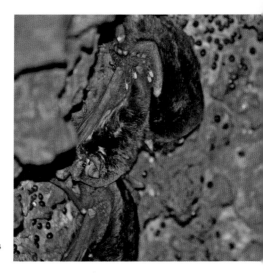

DESCRIPTION Small bat with average forearm length of 45mm. Characterized by enlarged second phalanx of third finger, which is folded within wing when the bat is at rest; hence 'bent-winged' (also called 'long-fingered bat'). Forehead rounded and elevated. Short and rhomboidal ears. Pelage soft and uniformly dark. Roosts gregariously in caves and tunnels. Females form maternity roosts while nursing pups. Flies out a little after dark. **DIET** Small beetles and flies. **DISTRIBUTION** Throughout India except Andaman and Nicobar Islands. **HABITAT** Feeds in open, uncluttered areas. Appears to favour hilly and forested country. Also shows a preference for streams and rivers.

Gangetic River Dolphin ▪ *Platanista gangetica*
F Platanistidae **O** Perissodactyla

DESCRIPTION Endangered freshwater dolphin. Large species growing to up to 2m from head to tail. Body greyish-brown. Long, toothed beak, very small, hump-like dorsal fin and short, broad flippers. This dolphin is blind and navigates and forages through echolocation. It is either solitary, or more commonly occurs in small groups. While hunting in parties the trails of the animals can be seen on the surface of the water, and individuals surface at regular intervals to breathe. **DIET** Fish, clams, shrimps and prawns. **DISTRIBUTION** Scattered locations in Ganga, Brahmaputra (and their tributaries) and Chambal; Beas and Harike (where referred to as Indus River Dolphin *P. gangetica minor*). **HABITAT** Slow-flowing parts of large rivers and beels (in Assam) with floating vegetation.

Indian Ocean Humpback Dolphin ▪ *Sousa plumbea*
F Delphinidae **O** Perissodactyla

DESCRIPTION Common dolphin along seashores in India. Measures 2–2.5m in length, and has a long, slender beak and small melon. Small but prominent falcate dorsal fin sits on a hump in middle of back; hump is more prominent in adults than in calves. Adults dark grey with pinkish pigmentation on the extremities, while calves are uniform grey. Surfaces more slowly and is less acrobatic than other species, such as Bottlenose Dolphin (see opposite), and adults rarely breach fully out of the water. Group sizes vary from solitary individuals and mother-calf pairs, to up to 20 individuals. **DIET** Fish (partial to mullets in estuaries), molluscs and crustaceans. **DISTRIBUTION** Throughout west and east coasts. Species has been split into **Indo-Pacific Humpback Dolphin** *S. chinensis* on east coast and **Indian Ocean Humpback Dolphin** *S. plumbea* on west coast. **HABITAT** Shallow waters (usually within depths of 20m) close to shores and tidal creeks; also enters rivers and estuaries.

Irrawady Dolphin ■ *Orcaella brevirostris*
F Delphinidae **O** Perissodactyla

DESCRIPTION Smaller than Indian Ocean Humpback Dolphin (see p. 154), with a large, bulging forehead (owing to large melon) and small, sickle-shaped dorsal fin at midback (in which it differs from superficially similar **Finless Porpoise** *Neophocaena phocaenoides*). Large flippers with curved leading edges. Bluish-grey in colour. Mouthline is horizontal. Feeds solitarily or in small groups – silently and inconspicuously – surfacing gently to blow. Also

has unique habit of 'spitting' jets of water; purpose of this is not clearly known, but some think that it is a means of stunning fish. **DIET** Fish, crustaceans and molluscs. **DISTRIBUTION** East coast mainly in Orissa, Andhra Pradesh and West Bengal (Sunderbans). **HABITAT** Shallow coastal waters, estuaries and mangrove creeks. Also Chilika lagoon.

Bottlenose Dolphin ■ *Tursiops truncatus*
F Delphinidae **O** Perissodactyla

DESCRIPTION Comprises two species, this one and **Indian Ocean Bottlenose Dolphin** *T. aduncus*. Large and robust dolphin, growing to up to 2–2.5m in length in Indian

waters; *T. aduncus* is smaller. Short beak, bulbous head with well-developed melon and moderately tall, sickle-shaped dorsal fin. Snout separated from forehead by crease. Colour generally dark grey on back and paler below. Rather acrobatic in the open ocean. Found in small groups, and sometimes associates with other whales and dolphins. **DIET** Bottom-dwelling, coastal and pelagic fish, rays, shrimps and crustaceans. **DISTRIBUTION** Throughout east and west coasts, and archipelagos. Some records from east coast refer to *T. aduncus*. Cosmopolitan distribution throughout the world. **HABITAT** Coastal as well as pelagic.

Pantropical Spotted Dolphin ■ *Stenella attenuata*
F Delphinidae O Perissodactyla

DESCRIPTION Elegant species, similar in size to Spinner Dolphin (see below). Slender in build with moderately long beak and flippers. Dorsal fin curved backwards. Distinct dark

grey cape that is high above flipper and dips just ahead of dorsal fin. Black line extends from eyes to beak, and bold black stripe runs from flipper to mouth. Body can be mottled, but not all individuals show mottling. Acrobatic like Spinner Dolphin, swimming and leaping out of the water repeatedly, but does not spin longitudinally as does Spinner. Lives in large pods, often with several hundred individuals. Tuna aggregate with these dolphins. **DIET** Small fish and squid. **DISTRIBUTION** Recorded from scattered locations off both east and west coasts, and Lakshadweep. **HABITAT** Open sea and near island reefs.

Spinner Dolphin ■ *Stenella longirostris*
F Delphinidae O Perissodactyla

DESCRIPTION Rather small (1.3–2m average length) and elegant species of dolphin. Slender in build with extremely long, thin beak, long and pointed flippers, and large, trianglular to slightly falcate dorsal fin. Melon and beak separated by crease. General body colour is a tripartite (three-toned) grey – darkest above, lighter along sides and pale below. Extremely acrobatic, swimming fast and leaping out of the water; the only dolphin that spins longitudinally several times before falling back into the water. Lives in large pods of 30 to several hundred individuals, often mixing with other species of cetacean and Yellow Fin Tuna. **DIET** Small fish and squid. **DISTRIBUTION** Recorded from scattered locations off both east and west coasts, and near Lakshadweep and Andaman-Nicobar archipelagos. **HABITAT** Open sea, continental slope and near island reefs.

Killer Whale (Orca) ▪ *Orcinus orca*
F Delphinidae **O** Perissodactyla

DESCRIPTION Handsome dolphin, unique in appearance. Largely black in colour with white ellipses behind eyes; grey saddle-patch behind dorsal fin; ventral white extends upwards to sides below and just behind dorsal fin. In females, dorsal fin is tall and falcate.

Adult males have erect, triangular dorsal fin, and flippers are also relatively large and rounded. Acrobatic, swims fast and often leaps out of the water. Lives in small, close-knit pods, and mostly seen as matrilineal groups. **DIET** Feeds exclusively on fish and in some places regularly hunts dolphins; also possesses ability to hunt down whale calves and adult whales. **DISTRIBUTION** Very few records exist from scattered locations off both the east and west coasts, and the archipelagos. Worldwide distribution. **HABITAT** Open sea but also comes close to coasts.

Sperm Whale ▪ *Physeter macrocephalus*
F Physeteridae **O** Perissodactyla

DESCRIPTION Huge whale with very large, distinctive square head, and rather narrow lower jaw. Body dark grey or brownish-grey. No dorsal fin. Series of bumps on dorsal ridge on tail stock, and flippers are also short and broad. Emits a distinctive forwards-angled, bushy blow through its single blowhole. The Sperm Whale's enormous rectangular head houses the largest brain in the Animal Kingdom. The species is also among the deepest divers. Mostly found in nursery schools of female with young or bachelor groups comprising several males. Produces very complex high-frequency vocalizations. **DIET** Deep-sea squid and fish. **DISTRIBUTION** Recorded a handful of times off the coasts of Gujarat, Karnataka, Maharashtra, Kerala, Lakshadweep islands, Tamil Nadu and Puducherry. Worldwide distribution. **HABITAT** Open sea and deep oceans.

Bryde's Whale ■ *Balaenoptera edeni*
F Balaenopteridae **O** Perissodactyla

DESCRIPTION Large animal growing up to 15–16m in length. Body generally smoky-grey in colour. Three prominent ridges on head ahead of blowholes are a distinctive feature. Blow can be columnar or bushy, and is highly variable. Dorsal fin is short (but tall in

comparison to dorsal fins of other whales), sickle shaped and pointed; it is located one-third of body from tail fluke. Found in groups of 5–6 individuals. **DIET** Schooling fish and zooplankton. **DISTRIBUTION** A few records from scattered locations off east and west coasts of Orissa, Tamil Nadu, Kerala, Karnataka, Goa, Lakshadweep and Maharashtra. Found chiefly in warm tropical and subtropical waters. **HABITAT** Both coastal and offshore.

Humpback Whale ■ *Megaptera novaeangliae*
F Balaenopteridae **O** Perissodactyla

DESCRIPTION Large and robust whale with adults measuring 10–15m in length. Slender head, with top flattened and covered by a number of fleshy knobs. Dorsal ridge on midline of head is indistinct. Dorsal fin situated one-third of body length from tail fluke. Often lobtails before feeding, and breaches frequently. **DIET** Swarming crustaceans (mostly in temperate waters) and shoaling fish. **DISTRIBUTION** There is a small non-migratory population in the Arabian Sea that is infrequently sighted off the Gulf of Kutch; stranded whales have been found in Kerala and Tamil Nadu. Cosmopolitan distribution throughout the world. **HABITAT** Open sea and oceans.

Blue Whale ▪ *Balaenoptera musculus*
F Balaenopteridae **O** Perissodactyla

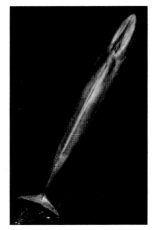

DESCRIPTION Largest animal in the world, growing to up to 24m in region covered by this book (subspecies Pygmy Blue Whale *B. m. brevicauda*). Body broad and head U shaped. Slim flippers are long and pointed, and dorsal fin is very small, positioned about one-third of body from tail fluke. Colour generally blue-grey, mottled with grey or white. Lives in groups of 3–4 individuals. Not known to dive deep in search of food. **DIET** Chiefly krill in temperate waters; diet unknown in tropical waters. **DISTRIBUTION** Recorded from scattered locations off east and west coasts, with several recent reports off Maharashtra and one off Karnataka. Regularly seen off north-eastern and southern coasts of Sri Lanka. **HABITAT** Open ocean, and some reports from coastal waters.

Dugong ▪ *Dugong dugon*
F Dugongidae **O** Sirenia

DESCRIPTION Large animal that grows to up to 3m in length. Similar in build to a dolphin, but appears more bulky and is sluggish and not adapted for swift motion. Face like that of a seal with a broad muzzle. No dorsal fin, and flippers are short, broad and rounded. Generally grey in colour and often shows heavy scarring on back, some of which can be due to cuts by propellers of boats. Chiefly solitary, grazing extensively on seagrass meadows; devours these from the base, hence leaving behind telltale signs of its presence. **DIET** Seagrass. **DISTRIBITION** Historically common in Gulf of Kutch and Gulf of Mannar, but now largely locally extinct with very occasional records. Currently, Andaman and Nicobar Islands seem to be its last fragile stronghold in this region. **HABITAT** Seagrass meadows in shallow waters.

IUCN RED LIST STATUS

CR	critically endangered	NT	near-threatened
EN	endangered	LC	least concern
DD	conservation dependant	VU	vulnerable

Common English Name	Scientific Name	Family	IUCN
Primata (Primates)			
Western Hoolock Gibbon	*Hoolock hoolock*	Hylobatidae	EN
Eastern Hoolock Gibbon	*Hoolock leuconedys*	Hylobatidae	VU
Bengal Slow Loris	*Nycticebus bengalensis*	Lorisidae	VU
Slender Loris	*Loris lydekkerianus*	Lorisidae	LC
Rhesus Macaque	*Macaca mulatta*	Cercopithecidae	LC
Bonnet Macaque	*Macaca radiata*	Cercopithecidae	LC
Arunachal Macaque	*Macaca munzala*	Cercopithecidae	EN
Assamese Macaque	*Macaca assamensis*	Cercopithecidae	NT
Long-tailed Macaque	*Macaca fascicularis*	Cercopithecidae	LC
Lion-tailed Macaque	*Macaca silenus*	Cercopithecidae	EN
Northern Pig-tailed Macaque	*Macaca leonina*	Cercopithecidae	VU
Stump-tailed Macaque	*Macaca arctoides*	Cercopithecidae	VU
White-cheeked Macaque	*Macaca leucogenys*	Cercopithecidae	-
Northern Plains Langur	*Semnopithecus entellus*	Cercopithecidae	LC
South-western Langur	*Semnopithecus hypoleucos*	Cercopithecidae	VU
Tufted Grey Langur	*Semnopithecus priam*	Cercopithecidae	NT
Terai Langur	*Semnopithecus hector*	Cercopithecidae	NT
Himalayan Langur	*Semnopithecus schistaceus*	Cercopithecidae	LC
Kashmir Grey Langur	*Semnopithecus ajax*	Cercopithecidae	EN
Nilgiri Langur	*Semnopithecus johnii*	Cercopithecidae	VU
Capped Langur	*Trachypithecus pileatus*	Cercopithecidae	VU
Golden Langur	*Trachypithecus geei*	Cercopithecidae	EN
Phayre's Leaf Monkey	*Trachypithecus phayrei*	Cercopithecidae	EN
Proboscidea (Elephants)			
Asiatic Elephant	*Elephas maximas*	Elephantidae	EN
Perissodactyla (Odd-toed hoofed mammals)			
Asiatic Wild Ass	*Equus hemionus*	Equidae	NT
Tibetan Wild Ass	*Equus kiang*	Equidae	LC

Indian Rhinoceros	*Rhinoceros unicornis*	Rhinocerotidae	VU
Artiodactyla (Even-toed hoofed mammals and cetaceans)			
Indian Chevrotain	*Moschiola indica*	Tragulidae	LC
Himalayan Musk Deer	*Moschus leucogaster*	Moschidae	EN
Alpine Musk Deer	*Moschus chrysogaster*	Moschidae	EN
Kashmir Musk Deer	*Moschus cupreus*	Moschidae	EN
Black Musk Deer	*Moschus fuscus*	Moschidae	EN
Indian Muntjac	*Muntiacus muntjak*	Cervidae	LC
Sambar	*Rusa unicolor*	Cervidae	VU
Tarim Red Deer (Hangul)	*Cervus hanglu*	Cervidae	LC (regionally threatened)
Brow-antlered Deer	*Rucervus eldii*	Cervidae	EN
Swamp Deer	*Rucervus duvaucelii*	Cervidae	VU
Spotted Deer	*Axis axis*	Cervidae	LC
Hog Deer	*Axis porcinus*	Cervidae	EN
Gaur	*Bos gaurus*	Bovidae	VU
Mithun	*Bos frontalis*	Bovidae	-
Wild Yak	*Bos mutus*	Bovidae	VU
Wild Buffalo	*Bubalus arnee*	Bovidae	EN
Nilgai	*Boselaphus tragocamelus*	Bovidae	LC
Four-horned Antelope	*Tetracerus quadricornis*	Bovidae	VU
Chinkara	*Gazella bennettii*	Bovidae	LC
Blackbuck	*Antilope cervicapra*	Bovidae	NT
Tibetan Antelope	*Pantholops hodgsonii*	Bovidae	EN
Tibetan Gazelle	*Procopra picticaudata*	Bovidae	NT
Asiatic Ibex	*Capra sibirica*	Bovidae	LC
Markhor	*Capra falconeri*	Bovidae	NT
Ladakh Urial	*Ovis orientalis*	Bovidae	VU
Argali	*Ovis ammon*	Bovidae	NT
Bharal	*Pseudois nayaur*	Bovidae	LC
Himalayan Tahr	*Hemitragus jemlahicus*	Bovidae	NT
Nilgiri Tahr	*Nilgiritragus hylocrius*	Bovidae	EN
Himalayan Brown Goral	*Naemorhaedus goral*	Bovidae	NT
Long-tailed Goral	*Naemorhedus caudatus*	Bovidae	VU
Red Goral	*Naemorhedus baileyi*	Bovidae	VU
Mishmi Takin	*Budorcas taxicolor*	Bovidae	VU

Himalayan Serow	*Capricornis thar*	Bovidae	NT
Red Serow	*Capricornis rubidus*	Bovidae	NT
Indian Wild Pig	*Sus scrofa*	Suidae	LC
Pygmy Hog	*Porcula salvania*	Suidae	CR
Cetacea (Whales, dolphins and porpoises)			
Gangetic River Dolphin	*Platanista gangetica*	Platanistidae	EN
Indo-Pacific Humpback Dolphin	*Sousa plumbea*	Delphinidae	NT
Finless Porpoise	*Neophocaena phocaenoides*	Phocoenidae	VU
Irrawady Dolphin	*Orcaella brevirostris*	Delphinidae	VU
Indo-Pacific Bottlenose Dolphin	*Tursiops aduncus*	Delphinidae	DD
Long-beaked Common Dolphin	*Delphinus capensis*	Delphinidae	DD
Long-snouted Spinner Dolphin	*Stenella longirostris*	Delphinidae	DD
Pan-tropical Spotted Dolphin	*Stenella attenuata*	Delphinidae	LC
Rough-toothed Dolphin	*Steno bredanensis*	Delphinidae	LC
Risso's Dolphin	*Grampus griseus*	Delphinidae	LC
Short-finned Pilot Whale	*Globicephala macrorhynchus*	Delphinidae	DD
Melon-headed Whale	*Peponocephala electra*	Delphinidae	LC
Killer Whale	*Orcinus orca*	Delphinidae	DD
False Killer Whale	*Pseudorca crassidens*	Delphinidae	DD
Sperm Whale	*Physeter macrocephalus*	Physeteridae	VU
Pygmy Sperm Whale	*Kogia breviceps*	Kogiidae	DD
Dwarf Sperm Whale	*Kogia sima*	Kogiidae	DD
Cuvier's Beaked Whale	*Ziphius cavirostris*	Ziphiidae	LC
Blainville's Beaked Whale	*Mesoplodon densirostris*	Ziphiidae	DD
Minke Whale	*Balaenoptera acutorostrata*	Balaenopteridae	LC
Humpback Whale	*Megaptera novaeangliae*	Balaenopteridae	LC
Sei Whale	*Balaenoptera borealis*	Balaenopteridae	EN
Bryde's Whale	*Balaenoptera edeni*	Balaenopteridae	DD
Fin Whale	*Balaenoptera physalus*	Balaenopteridae	EN
Blue Whale	*Balaenoptera musculus*	Balaenopteridae	EN
Carnivora (Carnivores)			
Tiger	*Panthera tigris*	Felidae	EN
Lion	*Panthera leo*	Felidae	VU
Leopard	*Panthera pardus*	Felidae	VU
Snow Leopard	*Uncia uncia*	Felidae	EN

Clouded Leopard	*Neofelis nebulosa*	Felidae	VU
Asian Golden Cat	*Catopuma temminckii*	Felidae	NT
Marbled Cat	*Pardofelis marmorata*	Felidae	NT
Caracal	*Caracal caracal*	Felidae	LC
Lynx	*Lynx lynx*	Felidae	LC
Pallas's Cat	*Otocolobus manul*	Felidae	NT
Jungle Cat	*Felis chaus*	Felidae	LC
Desert Cat	*Felis sylvestris*	Felidae	LC
Fishing Cat	*Prionailurus viverrinus*	Felidae	EN
Leopard Cat	*Prionailurus bengalensis*	Felidae	LC
Rusty-spotted Cat	*Prionailurus rubiginosus*	Felidae	NT
Spotted Linsang	*Prionodon pardicolor*	Prionodontidae	LC
Himalayan Palm Civet	*Paguma larvata*	Viverridae	LC
Common Palm Civet	*Paradoxurus hemaphroditus*	Viverridae	LC
Brown Palm Civet	*Paradoxurus jerdoni*	Viverridae	LC
Binturong	*Arctictis binturong*	Viverridae	VU
Small-toothed Palm Civet	*Arctogalidia trivirgata*	Viverridae	LC
Small Indian Civet	*Viverricula indica*	Viverridae	LC
Large Indian Civet	*Viverra zibetha*	Viverridae	LC
Malabar Civet	*Viverra civettina* *	Viverridae	CR
Grey Mongoose	*Herpestes edwardsii*	Herpestidae	LC
Ruddy Mongoose	*Herpestes smithii*	Herpestidae	LC
Small Indian Mongoose	*Herpestes auropunctatus*	Herpestidae	LC
Marsh Mongoose	*Herpestes palustris*	Herpestidae	-
Crab-eating Mongoose	*Herpestes urva*	Herpestidae	LC
Stripe-necked Mongoose	*Herpestes vitticollis*	Herpestidae	LC
Brown Mongoose	*Herpestes fuscus*	Herpestidae	LC
Striped Hyena	*Hyaena hyaena*	Hyaenidae	NT
Grey Wolf	*Canis lupus*	Canidae	LC
Jackal	*Canis aureus*	Canidae	LC
Asiatic Wild Dog	*Cuon alpinus*	Canidae	EN
Tibetan Sand Fox	*Vulpes ferrilata*	Canidae	LC
Indian Fox	*Vulpes bengalensis*	Canidae	LC
Red Fox	*Vulpes vulpes*	Canidae	LC
Desert Fox	*Vulpes pusilla*	Canidae	-

Asiatic Black Bear	*Ursus thibetanus*	Ursidae	VU
Himalayan Brown Bear	*Ursus arctos*	Ursidae	LC
Sloth Bear	*Melursus ursinus*	Ursidae	VU
Sun Bear	*Helarctos malayanus*	Ursidae	VU
Red Panda	*Ailurus fulgens*	Ursidae	EN
Small-toothed Ferret Badger	*Melogale moschata*	Mustelidae	LC
Large-toothed Ferret Badger	*Melogale personata*	Mustelidae	LC
Hog Badger	*Arctonyx collaris*	Mustelidae	VU
Honey Badger	*Mellivora capensis*	Mustelidae	LC
Stone Marten	*Martes foina*	Mustelidae	LC
Yellow-throated Marten	*Martes flavigula*	Mustelidae	LC
Nilgiri Marten	*Martes gwatkinsii*	Mustelidae	VU
Smooth-coated Otter	*Lutrogale perspicillata*	Mustelidae	VU
Eurasian Otter	*Lutra lutra*	Mustelidae	NT
Small-clawed Otter	*Aonyx cinereus*	Mustelidae	VU
Mountain Weasel	*Mustela altaica*	Mustelidae	NT
Himalayan Stoat (Ermine)	*Mustela erminea*	Mustelidae	LC
Siberian Weasel	*Mustela sibirica*	Mustelidae	LC
Yellow-bellied Weasel	*Mustela kathiah*	Mustelidae	LC
Back-striped Weasel	*Mustela strigidorsa*	Mustelidae	LC
Lagomorpha (Hares and pikas)			
Indian Hare	*Lepus nigricollis*	Leporidae	LC
Desert Hare	*Lepus tibetanus*	Leporidae	LC
Woolly Hare	*Lepus oiostolus*	Leporidae	LC
Cape Hare	*Lepus capensis*	Leporidae	LC
Hispid Hare	*Caprolagus hispidus*	Leporidae	EN
Royle's Pika	*Ochotona roylei*	Ochotonidae	LC
Large-eared Pika	*Ochotona macrotis*	Ochotonidae	LC
Plataeu Pika	*Ochotona curzoniae*	Ochotonidae	LC
Ladakh Pika	*Ochotona ladacensis*	Ochotonidae	LC
Moupin's Pika	*Ochotona thibetana*	Ochotonidae	LC
Forrest's Pika	*Ochotona forresti*	Ochotonidae	LC
Nubra Pika	*Ochotona nubrica*	Ochotonidae	LC
Sikkim Pika	*Ochotona sikimaria*	Ochotonidae	-
Pholidota (Pangolins)			

Chinese Pangolin	*Manis pentadactyla*	Manidae	CR
Indian Pangolin	*Manis crassicaudata*	Manidae	EN

Scandentia (Tree shrews)

Madras Tree Shrew	*Anathana ellioti*	Tupaiidae	LC
Malay Tree Shrew	*Tupaia belangeri*	Tupaiidae	LC
Nicobar Tree Shrew	*Tupaia nicobarica*	Tupaiidae	EN

Eulipotyphla (Shrews, moles and hedgehogs)

Assam Mole Shrew	*Anourosorex squamipes*	Soricidae	LC
Himalayan Water Shrew	*Chimarrogale himalayica*	Soricidae	LC
Andaman Spiny Shrew	*Crocidura hispida*	Soricidae	VU
Andaman White-toothed Shrew	*Crocidura andamanensis*	Soricidae	CR
Asian Grey Shrew	*Crocidura attenuata*	Soricidae	LC
Kashmir White-toothed Shrew	*Crocidura pullata*	Soricidae	DD
Gueldenstaedt's White-toothed Shrew	*Crocidura suaveolens*	Soricidae	LC
Horsfield's Shrew	*Crocidura horsfieldii*	Soricidae	DD
Jenkin's Andaman Spiny Shrew	*Crocidura jenkinsi*	Soricidae	CR
Nicobar Spiny Shrew	*Crocidura nicobarica*	Soricidae	CR
Pale Grey Shrew	*Crocidura pergrisea*	Soricidae	DD
Zarudny's Rock Shrew	*Crocidura zarudnyi*	Soricidae	LC
South-east Asian Shrew	*Crocidura fuliginosa*	Soricidae	LC
Blanford's Long-tailed Shrew	*Episoriculus macrurus*	Soricidae	LC
Hodgson's Brown-toothed Shrew	*Episoriculus caudatus*	Soricidae	LC
Kelaart's Long-clawed Shrew	*Feroculus feroculus*	Soricidae	EN
Elegant Water Shrew	*Nectogale elegans*	Soricidae	LC
Eurasian Pygmy Shrew	*Sorex minutus*	Soricidae	LC
Flat-headed Kashmir Shrew	*Sorex planiceps*	Soricidae	LC
Tibetan Shrew	*Sorex thibetanus*	Soricidae	DD
Indian Long-tailed Shrew	*Soriculus leucops*	Soricidae	LC
Sikkim Large-clawed Shrew	*Soriculus nigrescens*	Soricidae	LC
Anderson's Shrew	*Suncus stoliczkanus*	Soricidae	LC
Day's Shrew	*Suncus dayi*	Soricidae	EN
Hill Shrew	*Suncus niger*	Soricidae	EN
Grey Musk Shrew	*Suncus murinus*	Soricidae	LC
Pygmy White-toothed Shrew	*Suncus etruscus*	Soricidae	LC
Short-tailed Mole	*Euroscaptor micrura*	Talpidae	LC

White-tailed Mole	*Parascaptor leucura*	Talpidae	LC
Indian Hedgehog	*Paraechinus micropus*	Erinaceidae	LC
Collared (Desert) Hedgehog	*Hemiechinus collaris*	Erinaceidae	LC
Madras Hedgehog	*Paraechinus nudiventris*	Erinaceidae	LC

Rodentia (Porcupines,, squirrels, rats and mice)

Indian Crested Porcupine	*Hystrix indica*	Hystricidae	LC
Himalayan Crestless Porcupine	*Hystrix brachyura*	Hystricidae	LC
Asiatic Brush-tailed Porcupine	*Atherurus macrourus*	Hystricidae	LC
Himalayan Marmot	*Marmota himalayana*	Sciuridae	LC
Long-tailed Marmot	*Marmota caudate*	Sciuridae	LC
Indian Giant Squirrel	*Ratufa indica*	Sciuridae	NT
Malayan Giant Squirrel	*Ratufa bicolor*	Sciuridae	NT
Grizzled Giant Squirrel	*Ratufa macroura*	Sciuridae	NT
Red Giant Flying Squirrel	*Petaurista petaurista*	Sciuridae	LC
Indian Giant Flying Squirrel	*Petaurista philippensis*	Sciuridae	LC
Bhutan Giant Flying Squirrel	*Petaurista nobilis*	Sciuridae	NT
Woolly Flying Squirrel	*Eupetaurus cinereus*	Sciuridae	EN
Namdapha Flying Squirrel	*Biswamoyopterus biswasi*	Sciuridae	CR
Hodgson's Flying Squirrel	*Petaurista magnificus*	Sciuridae	LC
Spotted Giant Flying Squirrel	*Petaurista elegans*	Sciuridae	LC
Grey-headed Flying Squirrel	*Petaurista caniceps*	Sciuridae	LC
Kashmir Flying Squirrel	*Eoglaucomys fimbriatus*	Sciuridae	LC
Travancore Flying Squirrel	*Petinomys fuscocapillus*	Sciuridae	NT
Particoloured Flying Squirrel	*Hylopetes alboniger*	Sciuridae	LC
Hairy-footed Flying Squirrel	*Belomys pearsonii*	Sciuridae	DD
Hoary-bellied Squirrel	*Callosciurus pygerythrus*	Sciuridae	LC
Pallas's Squirrel	*Callosciurus erythraeus*	Sciuridae	LC
Himalayan Striped Squirrel	*Tamiops macclellandi*	Sciuridae	LC
Orange-bellied Himalayan Squirrel	*Dremomys lokriah*	Sciuridae	LC
Perny's Long-nosed Squirrel	*Dremomys pernyi*	Sciuridae	LC
Asian Red-cheeked Squirrel	*Dremomys rufigenis*	Sciuridae	LC
Three-striped Palm Squirrel	*Funambulus palmarum*	Sciuridae	LC
Five-striped Palm Squirrel	*Funambulus pennantii*	Sciuridae	LC
Dusky Striped Squirrel	*Funambulus sublineatus*	Sciuridae	VU
Jungle Striped Squirrel	*Funambulus tristriatus*	Sciuridae	LC

Indian Gerbil	Tatera indica	Muridae	LC
Indian Desert Jird	Meriones hurrianae	Muridae	LC
Little Indian Hairy-footed Gerbil	Gerbillus gleadowi	Muridae	LC
Pygmy Gerbil	Gerbillus nanus	Muridae	LC
Indian Long-tailed Tree Mouse	Vandeleuria oleracea	Muridae	LC
Nilgiri Long-tailed Mouse	Vandeleuria nilagirica	Muridae	EN
Long-tailed Field Mouse	Apodemus sylvaticus	Muridae	LC
Fukien Wood Mouse	Apodemus draco	Muridae	LC
Miller's Wood Mouse	Apodemus rusiges	Muridae	LC
Himalayan Wood Mouse	Apodemus pallipes	Muridae	LC
Pencil-tailed Tree Mouse	Chiropodomys gliroides	Muridae	LC
House Mouse	Mus musculus	Muridae	LC
Little Indian Field Mouse	Mus booduga	Muridae	LC
Bonhote's Mouse	Mus famulus	Muridae	EN
Spiny Field Mouse	Mus platythrix	Muridae	LC
Fawn-coloured Mouse	Mus cervicolor	Muridae	LC
Wroughton's Mouse	Mus phillipsi	Muridae	LC
Sikkim Mouse	Mus pahari	Muridae	LC
Harvest Mouse	Micromys minutus	Muridae	LC
Elliot's Brown Spiny Mouse	Mus saxicola	Muridae	LC
Pygmy Field Mouse	Mus terricolor	Muridae	LC
Cooke's Mouse	Mus cookii	Muridae	LC
Crump's Mouse	Diomys crumpi	Muridae	DD
Soft-furred Field Rat	Millardia meltada	Muridae	LC
Kondana Rat	Millardia kondana	Muridae	CR
Sand-coloured Rat	Millardia gleadowi	Muridae	LC
White-tailed Wood Rat	Madromys blanfordi	Muridae	LC
Indian Bush Rat	Golunda ellioti	Muridae	LC
Kutch Rock Rat	Cremnomys cutchicus	Muridae	LC
Ellerman's Rat	Cremnomys elvira	Muridae	CR
Manipur Rat	Berylmys manipulus	Muridae	DD
Chestnut Rat	Niviventer fulvescens	Muridae	LC
Smoke-bellied Rat	Niviventer eha	Muridae	LC
Mishmi Rat	Niviventer brahma	Muridae	LC
White-bellied Rat	Niviventer niviventer	Muridae	LC

Lang Bian Rat	*Niviventer langbianis*	Muridae	LC
Tennaserim Rat	*Niviventer tenaster*	Muridae	LC
Confucian Rat	*Niviventer confucianus*	Muridae	LC
Bower's Rat	*Berylmys bowersi*	Muridae	LC
Kenneth's White-toothed Rat	*Berylmys mackenziei*	Muridae	DD
Millard's Large-toothed Rat	*Dacnomys millardi*	Muridae	DD
Hume's Manipur Bush Rat	*Hadromys humei*	Muridae	EN
Edward's Noisy Rat	*Leopoldamys edwardsi*	Muridae	LC
Large Bandicoot Rat	*Bandicota indica*	Muridae	LC
Lesser Bandicoot Rat	*Bandicota bengalensis*	Muridae	LC
Short-tailed Bandicoot Rat	*Nesokia indica*	Muridae	LC
Black Rat	*Rattus rattus*	Muridae	LC
Brown Rat	*Rattus norvegicus*	Muridae	LC
White-footed Himalayan Rat	*Rattus nitidus*	Muridae	LC
Himalayan Rat	*Rattus pyctoris*	Muridae	LC
Indo-Chinese Forest Rat	*Rattus andamanensis*	Muridae	LC
Ranjini's Rat	*Rattus ranjiniae*	Muridae	EN
Sahyadri Forest Rat	*Rattus satarae*	Muridae	VU
Oriental House Rat	*Rattus tanezumi*	Muridae	LC
Miller's Nicobar Rat	*Rattus burrus*	Muridae	
Andamans Archipelago Rat	*Rattus stoicus*	Muridae	VU
Car Nicobar Rat	*Rattus palmarum*	Muridae	VU
Malabar Spiny Dormouse	*Platacanthanomys lasiurus*	Platacanthomyidae	VU
Kashmir Birch Mouse	*Sicista concolor*	Sminthidae	LC
Bay Bamboo Rat	*Cannomys badius*	Spalacidae	LC
Hoary Bamboo Rat	*Rhizomys pruinosus*	Spalacidae	LC
Silvery Mountain Vole	*Alticola argentatus*	Cricetidae	LC
Stoliczka's Mountain Vole	*Alticola stoliczkanus*	Cricetidae	LC
Royle's Mountain Vole	*Alticola roylei*	Cricetidae	NT
Kashmir Mountain Vole	*Alticola montosa*	Cricetidae	VU
White-tailed Mountain Vole	*Alticola albicauda*	Cricetidae	DD
Sub-alpine Kashmir Vole	*Hyperacrius fertilis*	Cricetidae	NT
Coniferous Kashmir Vole	*Hyperacrius wynnei*	Cricetidae	LC
Sikkim Mountain Vole	*Neodon sikimensis*	Cricetidae	LC
Pere David's Red-backed Vole	*Eothenomys melanogaster*	Cricetidae	LC

Blyth's Mountain Vole	*Phaiomys leucurus*	Cricetidae	LC
Grey Hamster	*Cricetulus migratorius*	Cricetidae	LC
Ladakh Hamster	*Cricetulus alticola*	Cricetidae	LC
Desert Hamster	*Phodopus roborovskii*	Cricetidae	LC
Chiroptera (Bats)			
Indian Flying Fox	*Pteropus medius*	Pteropodidae	LC
Blyth's Flying Fox	*Pteropus melanotus*	Pteropodidae	VU
Island Flying Fox	*Pteropus hypomelanus*	Pteropodidae	LC
Large Flying Fox	*Pteropus vampyrus*	Pteropodidae	NT
Nicobar Flying Fox	*Pteropus faunulus*	Pteropodidae	VU
Fulvous Fruit Bat	*Rousettus leschenaultii*	Pteropodidae	LC
Greater Short-nosed Fruit Bat	*Cynopterus sphinx*	Pteropodidae	LC
Lesser Short-nosed Fruit Bat	*Cynopterus brachyotis*	Pteropodidae	LC
Cave Nectar Bat (Dawn Bat)	*Eonycteris spelaea*	Pteropodidae	LC
Salim Ali's Fruit Bat	*Latidens salimalii*	Pteropodidae	EN
Blanford's Fruit Bat	*Sphaerias blanfordi*	Pteropodidae	LC
Ratnaworabhan's Fruit Bat	*Megaerops niphanae*	Pteropodidae	LC
Greater Long-nosed Fruit Bat	*Macroglossus sobrinus*	Pteropodidae	LC
Greater Mouse-tailed Bat	*Rhinopoma microphyllum*	Rhinopomatidae	LC
Lesser Mouse-tailed Bat	*Rhinopoma hardwickii*	Rhinopomatidae	LC
Small Mouse-tailed Bat	*Rhinopoma muscatellum*	Rhinopomatidae	LC
Egyptian Free-tailed Bat	*Tadarida aegyptiaca*	Molossidae	LC
European Free-tailed Bat	*Tadarida teniotis*	Molossidae	LC
Wrinkle-lipped Bat	*Chaerephon plicatus*	Molossidae	LC
Wroughton's Free-tailed Bat	*Otomops wroughtonii*	Molossidae	DD
Black-bearded Tomb Bat	*Taphozous melanopogon*	Emballonuridae	LC
Long-winged Tomb Bat	*Taphozous longimanus*	Emballonuridae	LC
Naked-rumped Tomb Bat	*Taphozous nudiventris*	Emballonuridae	LC
Egyptian Tomb Bat	*Taphozous perforatus*	Emballonuridae	LC
Theobald's Tomb Bat	*Taphozous theobaldi*	Emballonuridae	LC
Pouch-bearing Bat	*Saccolaimus saccolaimus*	Emballonuridae	LC
Greater False Vampire	*Megaderma lyra*	Megadermatidae	LC
Lesser False Vampire	*Megaderma spasma*	Megadermatidae	LC
Greater Horseshoe Bat	*Rhinolophus ferrumequinum*	Rhinolophidae	LC
Rufous Horseshoe Bat	*Rhinolophus rouxii*	Rhinolophidae	LC

Intermediate Horseshoe Bat	*Rhinolophus affinis*	Rhinolophidae	LC
Chinese Rufous Horseshoe Bat	*Rhinolophus sinicus*	Rhinolophidae	LC
Lesser Horseshoe Bat	*Rhinolophus hipposideros*	Rhinolophidae	LC
Blyth's Horseshoe Bat	*Rhinolophus lepidus*	Rhinolophidae	LC
Least Horseshoe Bat	*Rhinolophus pusillus*	Rhinolophidae	LC
Little Nepalese Horseshoe Bat	*Rhinolophus subbadius*	Rhinolophidae	LC
Shortridge's Horseshoe Bat	*Rhinolophus shortridgei*	Rhinolophidae	LC
Andaman Horseshoe Bat	*Rhinolophus cognatus*	Rhinolophidae	EN
Big-eared Horseshoe Bat	*Rhinolophus macrotis*	Rhinolophidae	LC
Thai Horseshoe Bat	*Rhinolophus siamensis*	Rhinolophidae	LC
Lesser Woolly Horseshoe Bat	*Rhinolophus beddomei*	Rhinolophidae	LC
Woolly Horseshoe Bat	*Rhinolophus luctus*	Rhinolophidae	LC
Pearson's Horseshoe Bat	*Rhinolophus pearsonii*	Rhinolophidae	LC
Dobson's Horseshoe Bat	*Rhinolophus yunanensis*	Rhinolophidae	LC
Trefoil Horseshoe Bat	*Rhinolophus trifoliatus*	Rhinolophidae	LC
Mitred Horseshoe Bat	*Rhinolophus mitratus*	Rhinolophidae	DD
Dusky Leaf-nosed Bat	*Hipposideros ater*	Hipposideridae	LC
Fulvous Leaf-nosed Bat	*Hipposideros fulvus*	Hipposideridae	LC
Andersen's Leaf-nosed Bat	*Hipposideros pomona*	Hipposideridae	LC
Least Leaf-nosed Bat	*Hipposideros cineraceus*	Hipposideridae	LC
Khajuria's Leaf-nosed Bat	*Hipposideros durgadasi*	Hipposideridae	EN
Kolar Leaf-nosed Bat	*Hipposideros hypophyllus*	Hipposideridae	EN
Nicobar Leaf-nosed Bat	*Hipposideros nicobarule*	Hipposideridae	-
Schneider's Leaf-nosed Bat	*Hipposideros speoris*	Hipposideridae	LC
Cantor's Leaf-nosed Bat	*Hipposideros galeritus*	Hipposideridae	LC
Great Himalayan Leaf-nosed Bat	*Hipposideros armiger*	Hipposideridae	LC
Horsfield's Leaf-nosed Bat	*Hipposideros larvatus*	Hipposideridae	LC
Kelaart's Leaf-nosed Bat	*Hipposideros lankadiva*	Hipposideridae	LC
Diadem Leaf-nosed Bat	*Hipposideros diadema*	Hipposideridae	LC
Tailless Leaf-nosed Bat	*Coelops frithii*	Hipposideridae	LC
Geoffrey's Trident Bat	*Asellia tridens*	Hipposideridae	LC
Stoliczka's Trident Bat	*Aselliscus stoliczkanus*	Hipposideridae	LC
Hodgson's Bat	*Myotis formosus*	Vespertilionidae	LC
Nepalese Whiskered Bat	*Myotis muricola*	Vespertilionidae	LC
Lesser Mouse-eared Bat	*Myotis blythii*	Vespertilionidae	LC

Whiskered Bat	*Myotis mystacinus*	Vespertilionidae	LC
Siliguri Bat	*Myotis siligorensis*	Vespertilionidae	LC
Mandelli's Mouse-eared Bat	*Myotis sicarius*	Vespertilionidae	VU
Kashmir Cave Bat	*Myotis longipes*	Vespertilionidae	DD
Hasselt's Bat	*Myotis hasseltii*	Vespertilionidae	LC
Horsfield's Bat	*Myotis horsfieldii*	Vespertilionidae	LC
Daubenton's Bat	*Myotis daubentonii*	Vespertilionidae	LC
Hairy-faced Bat	*Myotis annectans*	Vespertilionidae	LC
Peyton's Whiskered Bat	*Myotis peytoni*	Vespertilionidae	-
Common Pipistrelle	*Pipistrellus pipistrellus*	Vespertilionidae	LC
Indian Pipistrelle	*Pipistrellus coromandra*	Vespertilionidae	LC
Indian Pygmy Bat	*Pipistrellus tenuis*	Vespertilionidae	LC
Javan Pipistrelle	*Pipistrellus javanicus*	Vespertilionidae	LC
Mount Popa Pipistrelle	*Pipistrellus paterculus*	Vespertilionidae	LC
Japanese Pipistrelle	*Pipistrellus abramus*	Vespertilionidae	LC
Kuhl's Pipistrelle	*Pipistrellus kuhlii*	Vespertilionidae	LC
Kelaart's Pipistrelle	*Pipistrellus ceylonicus*	Vespertilionidae	LC
Black-gilded Pipistrelle	*Aerielulus circumdatus*	Vespertilionidae	LC
Chocolate Pipistrelle	*Falsitrellus affinis*	Vespertilionidae	LC
Savi's Pipistrelle	*Hypsugo savii*	Vespertilionidae	LC
Thomas's Pipistrelle	*Hypsugo cadornae*	Vespertilionidae	LC
Joffre's Pipistrelle	*Microstrellus joffrei*	Vespertilionidae	DD
Dormer's Bat	*Scotozous dormeri*	Vespertilionidae	LC
Great Evening Bat	*Ia io*	Vespertilionidae	LC
Common Serotine	*Eptesicus serotinus*	Vespertilionidae	LC
Bobrinskii's Serotine	*Eptesicus gobiensis*	Vespertilionidae	LC
Sombre Serotine	*Eptesicus tatei*	Vespertilionidae	DD
Botta's Serotine	*Eptesicus bottae*	Vespertilionidae	LC
Thick-eared Serotine	*Eptesicus pachyotis*	Vespertilionidae	LC
Parti-coloured Bat	*Vespertilio murinus*	Vespertilionidae	LC
Greater Yellow House Bat	*Scotophilus heathii*	Vespertilionidae	LC
Lesser Yellow House Bat	*Scotophilus kuhlii*	Vespertilionidae	LC
Yellow Desert Bat	*Scotoecus pallidus*	Vespertilionidae	LC
Hairy-winged Bat	*Harpiocephalus harpia*	Vespertilionidae	LC
Harlequin Bat	*Scotomanes ornatus*	Vespertilionidae	LC

Tickell's Bat	Hesperoptenus tickelli	Vespertilionidae	LC
Painted Bat	Kerivoula picta	Vespertilionidae	LC
Hardwicke's Woolly Bat	Keirvoula hardwickii	Vespertilionidae	LC
Papillose Bat	Kerivoula papillosa	Vespertilionidae	LC
Lenis Woolly Bat	Kerivoula lenis	Vespertilionidae	LC
Kachin Woolly Bat	Kerivoula kachinensis	Vespertilionidae	LC
Eastern Barbastelle	Barbastella darjelingensis	Vespertilionidae	LC
Lesser Flat-headed Bat	Tylonycteris pachypus	Vespertilionidae	LC
Greater Flat-headed Bat	Tylonycteris robustula	Vespertilionidae	LC
Common Noctule	Nyctalus noctula	Vespertilionidae	LC
Lesser Noctule	Nyctalus leisleri	Vespertilionidae	LC
Mountain Noctule	Nyctalus montanus	Vespertilionidae	LC
Hemprich's Long-eared Bat	Otonycteris hemprichii	Vespertilionidae	LC
Himalayan Long-eared Bat	Plecotus homochrous	Vespertilionidae	-
Kashmir Long-eared Bat	Plecotus wardii	Vespertilionidae	-
Round-eared Tube-nosed Bat	Murina cyclotis	Vespertilionidae	LC
Scully's Tube-nosed Bat	Murina tubinaris	Vespertilionidae	LC
Greater Tube-nosed Bat	Murina leucogaster	Vespertilionidae	DD
Hutton's Tube-nosed Bat	Murina huttoni	Vespertilionidae	LC
Little Tube-nosed Bat	Murina aurata	Vespertilionidae	LC
Rainforest Tube-nosed Bat	Murina pluvialis	Vespertilionidae	-
Jaintia Tube-nosed Bat	Murina jaintiana	Vespertilionidae	-
Peter's Tube-nosed Bat	Harpiola grisea	Vespertilionidae	DD
Eastern Bent-winged Bat	Miniopterus fuliginosus	Miniopteridae	LC
Nicobar Bent-winged Bat	Miniopterus pusillus	Miniopteridae	LC
Sanborn's Bent-winged Bat	Miniopterus magnater	Miniopteridae	LC
Sirenia (Dugongs)			
Dugong	Dugong dugon	Dugongidae	VU

■ FURTHER READING ■

Alempath, M. & Rice, C., *Nilgiritragus Hylocrius*, IUCN Red List of Threatened Species, Version 2013 (2008): 2.

Alfred, J. R., Ramakrishna, B. & Pradhan, M.S., 'Validation of Threatened Mammals of India', *Zoological Survey of India* (2006).

Bates, P. J. & Harrison, D.L., 'Bats of the Indian Subcontinent', Harrison Zoological Museum, 1997.

Choudhury, A., 'Primates in Northeast India: An Overview of their Distribution and Conservation Status', 2001, *ENVIS Bulletin: Wildlife & Protected Areas, Non-Human Primates of India*, Gupta A. K. (ed.), 1(1): 92–101.

Choudhury, A., *Mammals of Arunachal Pradesh*, Daya Books, 2003.

Corbett, G. B. & Hill, J. E., *Mammals of the Indo-Malayan Region: A Systematic Review*, Oxford University Press, 1992.

De Silva Wijeyeratne, G., *A Naturalist's Guide to the Mammals of Sri Lanka*, John Beaufoy Publishing, 2020.

Duckworth, J. W., Lunde, D. & Molur. S., *Tamios maclellandi*, IUCN 2013 Red list of Threatened Species. Version 2013 (2008):1.

Finn, F., *Sterndale's Mammals of India*, Thacker, Spink & Co. 1929.

Johnsingh, A. J. T. & Manjrekar, N., *Mammals of South Asia*, vols 1 & 11, Universities Press, 2013, 2015.

Karanth, U., 'Tiger', in Johnsingh, A. J. T. & Manjrekar, N. (eds), *Mammals of South Asia*, Universities Press, 2013.

Lydekker, R., *The Game Animals of India, Burma, Malaya, and Tibet*, R. Ward, 1907.

Menon, V., *Field Guide to Indian Mammals*, Dorling Kindersly (India) Pvt. Ltd., 2003.

Molur, S., Nameer, P. O. & Walker, S., 'Report of the Workshop Conservation Assessment and Management Plan for Mammals of India (BCPP-Endangered Species Project', Zoo Outreach Organisation, CSSG, Coimbatore, 1998.

Mukherjee, S., in Johnsingh A. J .T. & Manjerekar, N. (eds), *Mammals of South Asia*, Universities Press (India), Hyderabad, 2012.

Pfister, O., *Birds and Mammals of Ladakh*, Oxford University Press, 2004.

Phillips, W. W. A., *Manual of the Mammals of Sri Lanka*, Wildlife and Nature Protection Society, Colombo, 1980.

Pocock, R. I., *Fauna of British India: Mammals*, vol. 2, Taylor and Francis Ltd., 1941.

Prater, S. H., *The Book of Indian Animals* (2nd ed), BNHS and Oxford University Press, 1971.

Ranjitsinh, M. K., *Beyond the Tiger Portrait of Asian Wildlife*, Brijbasi Publications, 1997.

Roberts, T. J., *Mammals of Pakistan*, Oxford University Press, 1997.

Schaller, G. B., *Mountain Monarchs*: Wild Sheep and Goats of the Himalaya, University of Chicago Press, 1977.

Schaller, G. B., *Wildlife of the Tibetan Steppe*, University of Chicago Press, 1998.

Smith, A. & Xie, Y., *The Mammals of China*, Princeton University Press, 2008.

Srivastava, A., *Primates of Northeast India*, Megadiversity Press, 1999.

ACKNOWLEDGEMENTS

The authors thank Abhishek Jamalabad, Abhishek Gulshan, Aditya Singh, Amano Samarpan, Amit Sharma, Anuroop Krishnan, Anwaruddin Choudhury, Aparajita Datta, Ashok Kumar Das, Ashok Kashyap, Ashwin H. P., Ayan Banerjee, Bhavita Toliya, Biswapriya Rahut, Biswaroop Satpati, Clement M. Francis, Dhanu Paran, Dibyendu Ash, Dilan Mandanna, Dipani Sutaria, Divya Panicker, Dushyant Parasher, Garima Bhatia, Gaurav Shirodkar, Giri Cavale, Gopinath Kollur, HRS Urs, Kalyan Varma, Kamal Purohit, Karthikeyan Srinivasan, Khushboo Sharma, Manjula Mathur, Manoj Kejriwal, N. A. Naseer, N. S. Dhingra, Nandini Velho, Narayan Sharma, Niazul H. Khan, Nikhil Devasar, Niranjan Sant, Nishma Dahal, Otto Pfister, Pallavi Ghaskadbi, Pipat Soisook, Pritha Dey, Rajesh Puttaswamaiah, Rama Neelamegam, Ranjan K. Das, Rohan Pandit, Roon Bhuyan, Sachin Rai, Samyak Kaninde, Sarwandeep Singh, Savio Fonseca, Sayam U. Chowdhvry, Soumyajit Nandy, Subrata Debata, Sujan Chatterjee, Sumathi Sekhar, Sumit Sen, Taksh Sangwan, Tatiana Petrova, Tharaka Kusuminda, Tripta Sood, Udayan Borthakur, Umesh Srinivasan, Urmil Jhaveri, Vaidehi Gunjal, Vardhan Patankar, Vijay Kurhade, Vijay Sardesai and Vivek Sinha. We thank Divya Panicker and Abhishek Jamalabad for reviewing the text on marine mammals and Pritha for help with writing.

▪ INDEX ▪

Actitis binturong 90
Ailurus fulgens 99
Alticola stoliczkanus 125
Andaman Horseshoe Bat 142
Anderson's Leaf-nosed Bat 145
Antilope cervicapra 58
Aonyx cinerea 102
Apodemus sylvaticus 128
Arctonyx collaris 100
Arunachal Macaque 24
Asian Elephant 32
Asian Golden Cat 82
Asiatic Black Bear 97
Asiatic Ibex 62
Asiatic Wild Ass 34
Asiatic Wild Dog (Dhole) 95
Assamese Macaque 23
Atherurus macrourus 115
Axis axis 46
Axis porcinus 45
Balaenoptera edeni 158
Balaenoptera musculus 159
Bandicota bengalensis 130
Bandicota indica 129
Bengal Slow Loris 21
Bharal 64
Bhutan Giant Flying Squirrel
    119
Big-eared Horseshoe Bat 142
Binturong 90
Black-bearded Tomb Bat 138
Blackbuck 58
Black Rat (House Rat) 131
Blue Whale 159
Blyth's Horseshoe Bat 141
Bonnet Macaque 23
Bos frontalis 50
Bos gaurus 48
Bos mutus 51
Boselaphus tragocamelus 54
Bottlenose Dolphin 155
Brow-antlered Deer 43
Brown Bear 98
Brown Palm Civet 89
Brown Rat 131
Brush-tailed Porcupine 115
Bryde's Whale 158
Bubalus arnee 52
Budorcas taxicolor 71
Callosciurus erythraeus 122
Callosciurus pygerythrus 121

Canis aureus 95
Canis lupus 94
Cantor's Leaf-nosed Bat 147
Capped Langur 30
Capra sibirica 62
Capricornis thar 71
Caprolagus hispidus 107
Caracal 83
Caracal caracal 83
Catopuma temminckii 82
Cave Nectar Bat 134
Cervus elaphus 42
Chaerephon plicata 136
Chinese Ferret-badger 103
Chinese Pangolin 110
Chinkara 57
Clouded Leopard 83
Collared Hedgehog 114
Common Palm Civet 88
Crab-eating Mongoose 93
Cuon alpinus 95
Cynopterus brachyotis 133
Cynopterus sphinx 133
Desert Cat 86
Dremomys lokriah 123
Dugong 159
Dugong dugon 159
Dusky Leaf-nosed Bat 144
Dusky Striped Squirrel 125
Eastern Bent-winged Bat 153
Eastern Hoolock Gibbon 20
Egyptian Free-tailed Bat 136
Elephas maximus 32
Eonycteris spelaea 134
Equus hemionus 34
Equus kiang 35
Eurasian Otter 103
Felis chaus 85
Felis sylvestris 86
Fishing Cat 86
Five-striped Palm Squirrel 124
Four-horned Antelope 56
Fulvous Fruit Bat 132
Fulvous Leaf-nosed Bat 145
Funambulus palmarum 123
Funambulus pennantii 124
Funambulus sublineatus 125
Funambulus tristriatus 124
Gangetic River Dolphin 154
Gaur 48
Gazella bennettii 57

Golden Langur 30
Gollunda ellioti 129
Goral 70
Great Himalayan Leaf-nosed
    Bat 147
Greater False Vampire 139
Greater Horseshoe Bat 140
Greater Mouse-tailed Bat 135
Greater Short-nosed Fruit
    Bat 133
Greater Yellow House Bat 152
Grey-headed Flying Squirrel
    120
Grey Mongoose 91
Grey Musk Shrew 112
Grizzled Giant Squirrel 118
Hangul 42
Hardwicke's Woolly Bat 153
Helarctos malayanus 99
Hemiechinus collaris 114
Hemitragus jemlahicus 66
Herpestes auropunctatus 92
Herpestes edwardsii 91
Herpestes palustris 92
Herpestes smithii 91
Herpestes urva 93
Herpestes vitticollis 93
Hesperoptenus tickelli 152
Himalayan Crestless Porcupine
    115
Himalayan Langur 29
Himalayan Marmot 116
Himalayan Musk Deer 38
Himalayan Palm Civet 88
Himalayan Serow 71
Himalayan Striped Squirrel 122
Himalayan Tahr 66
Hipposideros armiger 147
Hipposideros ater 144
Hipposideros fulvus 145
Hipposideros galeritus 147
Hipposideros lankadiva 148
Hipposideros larvatus 146
Hipposideros pomona 145
Hipposideros speoris 146
Hispid Hare 107
Hoary-bellied Squirrel 121
Hodgson's Bat 148
Hog Badger 100
Hog Deer 45
Honey Badger 100

Hoolock hoolock 20
Hoolock leuconedys 20
Horsfield's Bat 149
Horsfield's Leaf-nosed Bat 146
House Mouse 128
Humpback Whale 158
Hyaena hyaena 94
Hylopetes alboniger 120
Hystrix brachyura 115
Hystrix indica 114
Indian Bush Rat 129
Indian Chevrotain 38
Indian Desert Jird 126
Indian Flying Fox 132
Indian Fox 96
Indian Gerbil 126
Indian Giant Flying Squirrel
    119
Indian Giant Squirrel 117
Indian Hare 106
Indian Hedgehog (Pale
    Hedgehog) 113
Indian Muntjac 39
Indian Ocean Humpback
    Dolphin 154
Indian Pangolin 110
Indian Porcupine 114
Indian Pygmy Bat 151
Indian Rhinoceros 36
Indian Wolf 94
Irrawady Dolphin 155
Jackal 95
Kashmir Long-eared Bat 151
Kelaart's Pipistrelle 150
Jungle Cat 85
Jungle Striped Squirrel 124
Kelaart's Leaf-nosed Bat 148
Kerivoula hardwickii 153
Killer Whale (Orca) 157
Ladakh Pika 109
Ladakh Urial 63
Large Bandicoot Rat 129
Large-eared Pika 108
Large Indian Civet 90
Latidens salimalii 134
Leopard 78
Leopard Cat 87
Lepus nigricollis 106
Lepus oiostolus 107
Lesser Bandicoot Rat 130
Lesser False Vampire 140

Lesser Mouse-tailed Bat 135
Lesser Noctule 150
Lesser Short-nosed Fruit Bat
    133
Lesser Woolly Horseshoe Bat
    143
Lion 76
Lion-tailed Macaque 26
Long-tailed Field Mouse 128
Long-tailed Macaque 24
Long-tailed Marmot 116
Long-tailed Tree Mouse 127
Long-winged Tomb Bat 137
Loris lydekkerianus 21
Lutra lutra 103
Lutrogale perspicillata 102
Lynx 84
Lynx lynx 84
Macaca arctoides 25
Macaca assamensis 23
Macaca fascicularis 24
Macaca leonina 25
Macaca mulatta 22
Macaca munzala 24
Macaca radiata 23
Macaca silenus 26
Madras Hedgehog 113
Malabar Spiny Dormouse 127
Malayan Giant Squirrel 117
Malay Tree Shrew 112
Manis crassicaudata 110
Manis pentadactyla 110
Marbled Cat 82
Marmota caudata 116
Marmota himalayana 116
Marsh Mongoose 92
Martes flavigula 101
Martes gwatkinsi 101
Megaderma lyra 139
Megaderma spasma 140
Megaptera novaeangliae 158
Mellivora capensis 100
Melogale moschata 103
Melursus ursinus 98
Meriones hurrinae 126
Miniopterus fuliginosus 153
Mishmi Takin 71
Mithun 50
Moschiola indica 38
Moschus leucogaster 38
Mountain Weasel 104

Moupin's Pika 109
Muntiacus muntjak 39
Mus musculus 128
Mustela altaica 104
Mustela erminea 105
Mustela kathiah 105
Mustela sibirica 104
Myotis formosus 148
Myotis horsfieldii 149
Myotis muricola 149
Naked-rumped Tomb Bat 138
Nemorhaedus goral 70
Neofelis nebulosa 83
Nepalese Whiskered Bat 149
Nesokia indica 130
Nicobar Tree Shrew 111
Nilgai 54
Nilgiri Langur 26
Nilgiri Marten 101
Nilgiri Tahr 68
Nilgiritragus hylocrius 68
Northern Pig-tailed Macaque
    25
Northern Plains Langur 27
Nyctalus leisleri 150
Nycticebus bengalensis 21
Ochotona ladacensis 109
Ochotona macrotis 108
Ochotona roylei 108
Ochotona thibetana 109
Octocolobus manul 84
Orange-bellied Himalayan
    Squirrel 123
Orcaella brevirostris 155
Orcinus orca 157
Otomops wroughtonii 137
Ovis orientalis 63
Paguma larvata 88
Pallas's Cat 84
Pallas's Squirrel 122
Panthalops hodgsonii 60
Panthera leo 76
Panthera pardus 78
Panthera tigris 74
Pantropical Spotted Dolphin 156
Paradoxurus hermaphroditus 88
Paradoxurus jerdoni 89
Paraechinus micropus 113
Paraechinus nudiventris 113
Pardofelis marmorata 82

Parti-coloured Flying Squirrel 120
Pearson's Horseshoe Bat 143
Petaurista caniceps 120
Petaurista nobilis 119
Petaurista petaurista 118
Petaurista philippensis 119
Petinomys fuscocapillus 121
Phayre's Leaf Monkey 31
Physeter macrocephalus 157
Pipistrellus ceylonicus 150
Pipistrellus tenuis 151
Platacanthomys lasiurus 127
Platanista gangetica 154
Plecotus wardi 151
Porcula salvania 72
Pouch-bearing Bat 139
Prionailurus bengalensis 87
Prionailurus rubiginosus 87
Prionailurus viverrinus 86
Prionodon pardicolor 106
Procapra picticaudata 61
Pseudois nayaur 64
Pteropus giganteus 132
Purple-faced Langur 31
Pygmy Hog 72
Rattus norvegicus 131
Rattus rattus 131
Ratufa bicolor 117
Ratufa indica 117
Ratufa macroura 118
Recervus duvaucelii 44
Recervus eldii 43
Red Fox 97
Red Giant Flying Squirrel 118
Red Panda 99
Rhesus Macaque 22
Rhinoceros unicornis 36
Rhinolophus beddomei 143
Rhinolophus cognatus 142
Rhinolophus ferrumequinum 140
Rhinolophus luctus 144
Rhinolophus macrotis 142
Rhinolophus pearsonii 143
Rhinolophus rouxii 141
Rhinophus lepidus 141
Rhinopoma hardwickii 135
Rhinopoma microphyllum 135
Rousettus leschenaultii 132
Royle's Pika 108
Ruddy Mongoose 91

Rufous Horseshoe Bat 141
Rusa unicolor 40
Rusty-spotted Cat 87
Saccolaimus saccolaimus 139
Salim Ali's Fruit Bat 134
Sambar 40
Schneider's Leaf-nosed Bat 146
Scotophilus heathii 152
Semnopithecus entellus 27
Semnopithecus hector 29
Semnopithecus hypoleucos 28
Semnopithecus johnii 26
Semnopithecus priam 28
Semnopithecus schistaceus 29
Short-tailed Bandicoot Rat 130
Siberian Weasel 104
Slender Loris 21
Sloth Bear 98
Small Indian Civet 89
Small Indian Mongoose 92
Smooth-coated Otter 102
Snow Leopard 80
Sousa plumbea 154
South-western Langur 28
Sperm Whale 157
Spinner Dolphin 156
Spotted Deer 46
Spotted Linsang 106
Stenella attenuata 156
Stenella longirostris 156
Stoat (Ermine) 105
Stoliczka's Mountain Vole 125
Striped Hyena 94
Stripe-necked Mongoose 93
Stump-tailed Macaque 25
Sun Bear 99
Suncus murinus 112
Sus scrofa 73
Swamp Deer 44
Tadarida aegyptiaca 136
Tamiops macclellandi 122
Taphozous longimanus 137
Taphozous melanopogon 138
Taphozous nudiventris 138
Tatera indica 126
Terai Langur 29
Tetracerus quadricornis 56
Three-striped Palm Squirrel 123
Tibetan Antelope 60
Tibetan Gazelle 61

Tibetan Sand Fox 96
Tibetan Wild Ass 35
Tickell's Bat 152
Tiger 74
Trachypithecus geei 30
Trachypithecus phayrei 31
Trachypithecus pileatus 30
Trachypithecus vetulus 31
Travancore Flying Squirrel 121
Tufted Grey Langur 28
Tupaia belangeri 112
Tupaia nicobarica 111
Tursiops truncatus 155
Uncia uncia 80
Ursus arctos 98
Ursus thibetanus 97
Vandeleuria oleracea 127
Viverra zibetha 90
Viverricula indica 89
Vulpes bengalensis 96
Vulpes ferrilata 96
Vulpes vulpes 97
Western Hoolock Gibbon 20
Wild Buffalo 52
Wild Pig 73
Wild Yak 51
Woolly Hare 107
Woolly Horseshoe Bat 144
Wrinkle-lipped Bat 136
Wroughton's Free-tailed Bat 137
Yellow-bellied Weasel 105
Yellow-throated Marten 101